# The Return of Hephaistos

**RECONSTRUCTING THE FRAGMENTED MYTHOS
OF THE MAKERS**

## Cheryl De Ciantis

Kairios Press

TUCSON, ARIZONA

Cheryl De Ciantis/Kairios Press

Tucson/Arizona/USA

www.kairios.com/resources

Book Cover Design and Art/Cheryl De Ciantis

Author Photo/Kenton Hyatt

Ordering Information: info@kairios.com

The Return of Hephaistos/Cheryl De Ciantis —2nd ed.

ISBN: 978-0-9857379-4-8

*for Kenton, always*

# Contents

*A myth is something that never was, but is always happening.*

—JEAN HOUSTON

(after Sallust, 1st century BCE)

# Preface to the Present Edition

IN THE SPRING of 2000, I was sitting with my graduate cohort of about 20 people in a comfortable, large lecture hall. We were still getting to know the program, the campus, the staff and faculty, and each other. Having chosen to devote my graduate study to mythology did not mean I felt I actually knew much about it. Dr. Ginette Paris was lecturing in our introductory class in Greek mythology and began to speak about a god with whom I was utterly unfamiliar: Hephaistos, the divine blacksmith, god of fire and of technology, who, uniquely in the Olympian pantheon is a wounded god. Born clubfooted, his imperfection enraged his mother Hera, queen of the gods, who threw the weakling infant from Olympus into the sea, a cosmically great distance below, where he was rescued and fostered by sea-nymphs who hid him in their submarine cavern. There, he was tutored in the magical arts of metallurgy, which can fairly be said to be the rocket science of the ancient Mediterranean.

This crippled god has a distinctive gait. Instead of describing it, Dr. Paris suddenly and very unexpectedly came around the lectern to where several of us sat in twos and threes in the front row of seats, bent her back and slowly crab-walked across the floor in Hephaistos' circular, clubfoot-dragging gait. I felt as though I'd been hit by a thunderbolt. It literally took my breath away. In that moment I knew for a certainty what my dissertation topic would be.

This dissertation was completed and accepted in 2005 for my Ph.D. in Mythological Studies with Emphasis in Depth Psychology at Pacifica Graduate Institute.[1] I am publishing an e-book and print

---

[1] It remains available through UMI's ProQuest dissertation service (UMI Number 3173606), and a full PDF version is available online. This edition includes a newly-created Index and Foreword.

v

version now for a few important reasons. One impetus is that in 2018 this dissertation abstract was rated among the top-ranked selections of that year for inclusion in the peer-reviewed Leonardo Abstracts Service (LABS), "an evolving, comprehensive database of thesis abstracts on topics at the intersections between art, science and technology"[2] Leonardo is the journal of art and technology published by The MIT Press, which in 2018 celebrated its 50th anniversary. LABS was instituted in 2006, and it is a testament to the relevance of myth in understanding and advancing both the origins and future of technology that this thesis was selected for inclusion, several years after its completion.

A very great deal has changed since this dissertation was successfully defended at the beginning of 2005. Then, the invisible golden net cast by Hephaistos, strong enough to bind adulterous Aphrodite and Ares in flagrante delicto (which you will read about in this book), seemed an obvious analogy for the Internet. Social media was still growing; it is hard to remember the era of dial-up internet, Myspace and Ask Jeeves given the massive floods of Internet-disseminated energies that have been unleashed in global upheavals since that time. The momentous Arab Spring that began in 2011 and engulfed four countries was significantly energized by Twitter. The surprise outcome of the U.S. 2016 election was influenced by Russian actors spreading viral disinformation using fake identities constructed with the help of "big" data mined from Facebook users by the now infamous (and defunct) Cambridge Analytica. Hephaistos' other creations, described by Homer and other ancient authors writing 28 centuries ago—industrial-grade robots capable of working in his forge, and a biotech masterpiece: the first flesh-and-blood woman, Pandora—show that Hephaistos is surely the paradigmatic deity of artificial intelligence and biotechnologies. We are only now just beginning to see the daidalic—that is to say, game-changing—implications of technologies like CRISPR gene-editing. We are also beginning to more clearly see the implications of technology that has far outstripped meaningful ethics frameworks, for example the disclosure by Chinese authorities in 2019 that a rogue Chinese scientist had cloned human infants. Hephaistos and the Makers—including smith-gods like Ogun, who is still revered in West Africa and the African diaspora orisha religions—are

---

[2] https://www.leonardo.info/labs-2018

demanding our attention because they have much to tell us about new questions that have old origins. Much has changed; and much remains, mythically, the same. The Makers, who are imaginally ambidextrous, easily compass "both/and" propositions.

Though the Makers are not typically among the best known mythic figures, when one chooses to look, the Makers reveal themselves, however fleetingly, in myths and traditions that appear everywhere in the world. Depth psychologist James Hillman insisted that the core tenet of archetypal psychology is "stick to the image." When we stick to the "biography" of Hephaistos and the Makers, discernable in widely dispersed fragments found in Homer, Hesiod and many other sources, what we see is that the Makers are magicians. They transform matter and are the source of good things for material life on earth. Precisely because of this instrumentality and its game-changing power, the Makers, and their mortal avatars— technologists, artists, shamans—are held both in awe-ful reverence and persisting suspicion. Arthur C. Clarke's "Third Law" puts it succinctly: Any sufficiently advanced technology is indistinguishable from magic. Its corollary is that any sufficiently advanced, mortal Maker is subject to extraordinary internal tensions in the dance between confidence and hubris; and externally, between being perceived as advancing what is "good" for the *polis* (the political collective) on one hand, and promoting subversive and therefore "bad" art, ideas or products.

This book is intended to inform Makers about their mythic (and psychically, quite alive, active and powerful) lineage in such as way as, hopefully, will encourage them to ethically navigate the extremities and to find the sweet spot in the twisting channel of constructive instrumentality, at a time when our planet needs nothing less than transformative energy that will benefit all its residents. This book is also intended to inform those who love or who manage Makers.

I have taken the opportunity to correct very minor errors that have persisted in the text of this dissertation, but otherwise left it unchanged with the exception of adding an index for the reader's convenience.

Tucson, Arizona 2019

CHAPTER 1

# Introduction

IN BOOK 18 of the Iliad, Homer tells of a wondrous shield, made by a god for a half-divine warrior to wield on the battlefield of Troy. The massive shield is built of five layers of bronze, tin, gold, and silver. On it are depicted the earth, sky, and sea, the blazing sun and full moon, and the constellations of the Pleiades, the Hyades, Orion the warrior, and the Great Bear, also known as the Wagon that wheels around the sky.

The images of two cities are forged on the shield. In one, weddings are being celebrated. The brides are being brought forth from their bridal chambers amid feasting and youths dancing to harp and flute as women rush to their doors to stare in wonder at the sight. Here, where the people stream to the marketplace, a sudden quarrel has broken out over the blood-price for a murdered kinsman. A judge is called to adjudicate. The cheering crowd takes sides as the city elders form the sacred circle where heralds will be heard to plead the case. See: two bars of gold shine on the ground, a prize for the wisest speech.

In another city a besieging army holds a war council. There is division: should they plunder the city or share the wealth with its citizens? However, the people of the city have no plans for surrender. Instead, they are arming themselves to march toward their besiegers. Wives, children, and elders guard the city ramparts. Ares and Pallas Athena lead them. The brilliance of the gods' magnificent armor is picked out in burnished gold, placed there by the forge-god's hand.

The two armies clash along the river banks outside the city. Strife, Havoc, and Death join the fray, seizing bloodied fighters in their frightful talons and dragging off the dead.

Between the two cities the god forges fallow fields beside fields where plowmen labor and are refreshed with honeyed wine at the end of each long furrow. Even though it is modeled in gold, see how the earth appears black and rich: such is the wonder of this work. Look here: harvesters scythe ripe grain on a king's estate, others bind the sheaves while beneath a spreading oak the king's heralds lay a magnificent harvest feast for the midday meal. Now a rich vineyard is rendered in bright enamel, fenced in tin. In it, young, high-hearted harvesters dart through the rows of vines. Listen: a boy plucks a lyre and sings a dirge for the dying year as the harvesters beat out the time with dancing steps. Look now: drovers driving a herd along a stream are being attacked by lions that seize a bull, ripping its hide and devouring its guts. The dogs shrink back from the lions' rage. Now here: a meadow, where sweetly garlanded girls and beautiful, oiled boys with daggers on their belts dance rapturously as tumblers whirl and spring among them. And bounding the whole, the god renders Ocean's massive river, girdling the shield's rim.

The shield is a physical impossibility, a small space crammed with a cosmos full of activity, truly the work of a god. Homer's *ekphrasis*, or description of it, is also filled with impossibilities. A metal shield depicts narratives and figures which shift and play through time; even sounds are heard. The tone of the poet's ekphrasis is itself rapturous, almost ecstatic in its own rushing momentum as it follows the shield's creation in every part. The power and imagination of the god is great indeed, and in his inspired and artful ekphrasis, the poet claims a share of divinity for himself.

The shield's maker is Hephaistos, the Greek god of fire and the forge and, together with Athena, the god of art and technology. He is a metalworker, a maker of magical objects like the marriage necklace of Harmonia, which confers irresistible beauty, as well as the deadly thunderbolts of Zeus. He is the architect who built the glorious brazen mansions of the Olympian gods and he is honored in the Homeric Hymns for making the lives of humankind easier:

> They had before then been dwelling in caves on the mountains like beasts,

But now, knowing works through Hephaistos renowned for his skill, with ease
Till the year brings its end they live in comfort within their own homes.[3]

Hephaistos is the only cripple among the Olympian gods, and it is said he is the only Olympian who works.[4]

The Hephaistos myth is mentioned in relatively few extant classical sources. The lengthiest and most developed versions of parts of the god's story are found in Homer's *Iliad* and *Odyssey*. Based largely on Homer's depictions, Hephaistos has come to be imagined as the jealous, dwarfish, lame, ugly, unsuccessful husband of Aphrodite, who occasions the scornful laughter of the other Olympian gods. He is the introverted peacemaker who gentles the thunderous conflict between his parents Zeus and Hera. He is a dutiful son to Hera and he fulfills the imperious commands of Zeus, who represents the principle of order in the Homeric universe.

When the labyrinthine threads of mythic connection are traced *backward* from the time of Homer into earlier Greek and Mediterranean sources, mythemes are revealed that diverge from Homer's depiction of Hephaistos. The scant and scattered clues found in both pre-classical and classical texts citing traces of earlier traditions add layers of mystery to the god's story. Some of these texts are themselves fragmentary or, like Homer's texts, selective in regard to the origins, character and associations of the mythic image of Hephaistos. Spreading the net *sideways* into the mythologies of other cultures that contain images of divine blacksmiths and related, often shadowy, magical, chthonic figures will help to cast light on questions raised by lacunae in the Greek material. A comparison of differing versions and variants of the Hephaistos myth with structurally and imagistically related versions of non-Greek myths will show the Hephaistean mythos to be a richly textured net of archetypal themes which include magic, mystery rites, and knowledge of the secrets of Earth, learned from the Great Mother herself.

---

[3] "Hymn to Hephaistos," *The Homeric Hymns*, trans. Michael Crudden, Oxford World's Classics (Oxford: Oxford University Press, 2001), 84-85.

[4] Murray Stein, "Hephaistos: A Pattern of Introversion," in *Facing the Gods*, ed. James Hillman (Dallas: Spring. 1980), 67.

Just as searching backwards reveals mystery, searching *forward* from Greek times reveals voids and absences. The presence of the god is not easy to trace and his name seems all but forgotten. Yet, it becomes possible to clearly see the archetypal image of Hephaistos in two seemingly unrelated and quite contemporary cultural myths. One concerns attitudes toward art and the artist. Few people consciously claim the divine gift of their own creativity and those who do are very often viewed with an uneasy mixture of admiration and suspicion. The unique ambiguities of the god's power have been reduced by some depth-psychological writers to the image of the limping figure whose psychic mother-wound must be overcome in order for a healthy exercise of socially positive creativity to be achieved. This mythopoetic—i.e., "myth-making"—strand follows what will be seen to contain vestiges of very old fears of the "left-hand" or "sinister" aspects of the legendary and divine creative powers anciently associated with Hephaistos. Its more recent manifestations have involved a pathologizing of the artist, who, unless co-opted within the marketplace or by the ruling political current, is often marginalized, stigmatized, ignominiously cast from society's Olympus. When the artist finds a way to express, however subtly, the volcanic anger that seethes below the restraints imposed by ruling structures, he or she may even be imprisoned or killed.

Hephaistos can also be discerned in a second cultural myth that envisions a massive web of global technology, capitalism, and geopolitics. In this cultural myth, technology is seen almost to have a will of its own independent of its makers to produce materiél that is indifferent to its own destructive capability. Industries that make "good things for life" also cast visible and invisible pollutants into the biosphere at a scale which many believe will destroy Earth. In this cultural myth science is viewed ambivalently as both a potential transgressor against divine laws and as containing hope for bettering the condition of human life.

Somehow, the mythic figure of Hephaistos has come to stand for psychological and social forces that threaten the very continuance of life on Earth. This mythic identification, both implicit and explicit, is well-exemplified in recent fiction and cinema and in headlines and "sound bites." As these threads are followed in the course of this dissertation, it will become increasingly clear that, in view of the

4

modern representation of the Hephaistean archetype, the product of centuries of mythopoeticizing, something has been lost. The questions this dissertation will examine include: What has been lost, fragmented, or discarded? What are among the possible consequences of this fragmentation? Who and what would a "recovered" Hephaistos be? And, finally: Why does it matter?

## Greek Hephaistos: An Ambiguous Image

In his depictions in ancient Greek literature, the figure of Hephaistos embodies a number of ambivalences and ambiguities. Living among the perfect Olympian gods, Hephaistos, uniquely, is lame. He is sometimes even described as dwarfish, ugly. Yet, Homer shows him married to the most beautiful of goddesses, Aphrodite, in the *Odyssey* and to one of the Graces in the *Iliad*. Honored and sought-after for his skill and craft, Hephaistos also draws the laughter of the other Olympians. In a tale-within-a-tale in Homer's *Odyssey*, Hephaistos sets a snare to trap the adulterous lovers Aphrodite and Ares in his own bed, whereupon he angrily calls the Olympians to witness the grievous evidence of his cuckholding. Instead of condemning the miscreants, Apollo and Hermes exchange laughter as Hermes loudly says he would gladly take the place of Ares under the net with the naked Aphrodite. The laughter of Hermes and Apollo has been interpreted as impugning Hephaistos' sexual prowess. Yet, Hephaistos' domestic life as depicted by Homer in the *Iliad* is harmonious, and his wife, Charis, is the most gracious of helpmeets. Her name means simply "Grace" and denotes her as one of the three beautiful handmaidens of Aphrodite.

Some versions of Hephaistos' myth deny him offspring; others attribute various children to him, but not always through normal sexual union. Chief among these is Erichthonios, snake-tailed child of Gaia, venerated as ancestor by the citizens of Athens to prove their claim as sons of the very Athenian earth they sprang from, a people self-generated in that place and entitled to rule everything around them. The best-known story of his origin is that Erichthonios is conceived from semen Athena brushes from her thigh, the result of the intemperate admiration shown her by Hephaistos.

Does Hephaistos represent deformed sexuality? Or, on the other hand, is Hephaistos perhaps uniquely successful in marrying the most beautiful goddesses? What is the nature and import of this god's sexuality?

A second significant Hephaistean ambivalence concerns his relationship with his mother, Hera, and other mother figures in his myth, and the ambiguity of his fathering. Homer's texts imply that Hephaistos is the son of Zeus and Hera. However, according to Hesiod, Hera conceives him parthenogenically out of jealousy when Zeus births fully-armed Athena. The timing of these events is confusing, since Hephaistos is often represented in vase paintings as midwifing at Athena's birth by splitting open Zeus's head with his forge hammer. In the version of his parthenogenic birth, he is born crippled, and Hera, in her disgust, casts him from Olympus. He falls into the ocean and is rescued and raised by the sea nymphs Thetis and Eurynome. In their cavern under the sea Hephaistos is provided with a tutor, Kedalion, who teaches him metallurgy. When news of his prodigious talent for making beautiful objects reaches Olympus, Zeus commands his presence there. Hephaistos refuses, but sends a magnificent golden throne for Hera. When she sits in it, it traps her and flies up in the air, suspending her helplessly. No Olympian can remedy the situation. Ares is sent to retrieve the prankster and fails. Dionysos succeeds, no doubt through the use of flattery as well as strong wine. One of the favorite subjects of Greek potters of the sixth and fifth centuries BCE is Hephaistos arriving on Olympus on the back of a mule, drunk, accompanied by Dionysos and Silenos. Homer (who ignored or did not know of Hephaistos' mule ride) shows both Hephaistos' apparent loyalty and kindness to his mother (in Book 1 of the *Iliad*) as well as his still-fresh anger—he terms his mother a "bitch" (in Book 18). The tricky smith is depicted as harboring ambivalent feelings at the least. Not, however, toward his foster mother Thetis, at whose request he creates the magnificent shield for Achilles, its divine brilliance so stunning that Achilles' men cower at first sight of it.

Is Hephaistos a bastard or a legitimate child of Olympus? Is he mother-wounded and impotent, or many-mothered and creatively potent? What is the nature of his potency, and what does it have to do with the mystery of his birth and parentage?

A third Hephaistean ambivalence concerns his inclusion in the machinations and manipulations of Zeus and other gods in times of conflict. Though he supports his mother Hera's plot against Zeus on one occasion (and suffers being thrown from Olympus by the angry god as a result) he later counsels her to knuckle under, doing so in such a way as to diffuse tension by drawing the laughter, first nervous, then relieved, of the other gods, who are fearful of Zeus's rage. Aeschylus' portrayal of Hephaistos in *Prometheus Bound* shows him duly carrying out the orders of Zeus to firmly chain Prometheus to the rock where his liver will daily be torn by an eagle. He takes care to perform an efficient job, but he does so with reluctance. He wishes there were someone else would do the job. Aeschylus even puts a criticism of Zeus into the mouth of Hephaistos, who offers in explanation if not amelioration of the punishment of Prometheus that, "Who holds a power but newly gained is ever stern of mood."[5]

Whose side is he on? Is he on the side of Olympian power? Is he on the side of humanity? Does he care, when, in the *Iliad*, he is called out by Hera onto the Trojan field of battle to blow his powerful, withering fire on the river god Xanthus, whether the Achaeans or Trojans will be victorious? Or, is he simply performing a job of work in the same manner as any military contractor? This is the attitude that David L. Miller implies when he characterizes Hephaistos together with Hera as mythically emblematic of the "military-industrial complex."[6] Whose side is Hephaistos on?

A fourth Hephaistean ambivalence is his relationship to the family of Olympian gods and to legitimacy itself. Cast out by his mother at birth, he is a late arrival to Olympus. His abodes include the undersea cave of Thetis and Eurynome, where he learns his arts and Mount Olympus where he occupies his own godly mansion similar to those he has created for his brother and sister gods. Sometimes he resides under Mount Etna, hidden away in his inaccessible forge. Ugly, he creates beauty. He makes things both inanimate and animate. Certain

---

5 Aeschylus, Prometheus Bound, trans. E. H. Plumptre, Vol. 8, Part 4, The Harvard Classics (New York: P. F. Collier and Son, 1909–14), 42, Bartleby.com, 2001, accessed March 15, 2004, http://www.bartleby.com/8/4/.
6 David L. Miller, The New Polytheism: Rebirth of the Gods and Goddesses (New York: Harper, 1974), 66-67.

of these things are demanded of him by Zeus, including the flesh-and-blood first woman, Pandora, whom Hephaistos forms from earth and water and gives a "voice and strength."[7] He fashions the intelligent girls made of gold who assist him in his forge, bellows that do his bidding in the most exacting tasks, tripods that wheel themselves around Olympus where needed and wheel themselves back to the forge again. He is not only physically different from the rest of the "happy gods," who but rarely occupy themselves with making (with the possible exception of Athena), but according to Homer he has also felt "mortal pain."[8] When Zeus casts him from lofty Olympus for taking the part of Hera, he falls for a whole day. Rescued and nursed by the Sintians, a non-Greek people, ancient occupants of the Aegean island of Lemnos, he forms a strong bond with these mortals, who speak an outlandish tongue and who name their capital city in his honor and ever after celebrate his cult.

Lamed, wounded, vulnerable, once (twice?) discarded but also indispensable, the maker of all good things for gods and mortals alike as well as the maker of magical things. Who is he?

## *Looking Backward (and Sideways) from Homer: The Divine Blacksmith in Myth and Folklore*

Following threads of mythic connection *backward* from the time of Homer yields compelling, sometimes confusing, information. The ancient Hephaistean mythos is exceedingly richly faceted and many-sided. Hephaistos' ancient status, as a blacksmith god, god of applied arts, fire and burnt sacrifice, and his wounded condition all indicate membership in a specific set of divine lineages. The mentions of these aspects of this divinity, while myriad in classical texts, are also scant in substance and shadowy in implication. We know almost nothing about the actual cult of Hephaistos in ancient Greece, except for the outlines of the major Athenian festival he shared with his sister Athena, the Chalkeia, and the mystery cult centers of the Kabeiroi—sometimes known as the "sons of Hephaistos"—on Lemnos and at Samothrace and Thebes.[9] However, the Hephaistean

---

7 Hesiod, Theogony, trans. Apostolos N. Athanassakis (Baltimore: Johns Hopkins University Press, 1983), 62-63.

8 Homer, *The Iliad*, trans. Robert Fagles (New York: Penguin, 1990), 18:462.

mythos is enmeshed in a large and extensive net of myths and traditions of the blacksmith/shaman/artist gods—*making* gods—who figure significantly in mythology and folklore from Africa to Ireland. One of the aims of this dissertation is to place Hephaistos within this mythic lineage system.

Specific connections can be made between the Hephaistean mythos and a complex of primordial archetypal figures. These include the ancient blacksmith/shaman/artist gods (and goddesses) and what I will term the cohort of *chthonic phallic* deities (a term I borrow from Erich Neumann). All are associated with the powers of the earth, hence "chthonic," meaning of the earth; many are associated with mysteries dealing with the continuum of terrestrial life with the underworld and death, hence chthonic in its mystical sense. And they are phallic: the mythical, often incestuous sons-and-lovers of the Great Goddess, associated with magic, whose powers are both summoned and averted by means of phallic charms, images, amulets, and symbolic representations.

Among them are the Cyclops, Kabeiroi, Daktyloi, and Telchines, who have left their traces in traditions fleetingly described in texts such as the *History* of Herodotus and Pausanias' catalogue of Greek places and customs. Primordial, mysterious, and sometimes confusingly inter-identified, they have maintained a shadowy existence throughout history in various metamorphosed forms, including the ubiquitous dwarf of folklore and fairy tale. These creatures share particular qualities that place them in intimate relationship with the forge and fire, knowledge of the earth and its minerals and treasures, and the uncanny skill of artisans and artists. Their relationship to humankind is elusive, indirect, and marginal, and their powers are sometimes viewed as beneficial, sometimes malign. Residing both literally and figuratively "underground," they partake in qualities shared with the shaman: knowledge of the properties of natural as well as supernatural things. Further, receiving their magical skills from the feminine Earth, the chthonic phallic deities are nurturers and healers.

---

[9] Timothy Gantz, *Early Greek Myth: A Guide to Literary and Artistic Sources* (Baltimore: The Johns Hopkins University Press, 1993), 148.

Lotte Motz observes that the etymology of the Germanic term *smith* indicates it is not limited in meaning to metal working, but indicates making generally.[10] She ultimately suggests that the Greek *poieio*—meaning "to make" and the origin of the word *poetry*—is the term that applies.[11] The divine chthonic phallic cohort are artful makers and technicians of the sacred. The Dactyls (and/or the Kabeiroi) are inventors of poetic meter, and smiths in Northern European mythology are associated with poetry and bardic song; strangers near the abodes of the dwarf mountain smiths are traditionally warned by the rhythmic music of hammer on anvil, said to be the origin of poetry.

One of the signal features of this complex of inter-related images is physical disability or some physical characteristic or marking which separates the mythic maker from the rest of gods and humankind, like the round eye in the middle of the Cyclops' foreheads. Jung notes that "Ugliness and deformity are especially characteristic of those mysterious chthonic gods, the sons of Hephaistos, the [Kabeiroi], to whom mighty wonder-working powers were ascribed." Jung notes also that "Hephaistos, Wieland the Smith and Mani (founder of Manichaeism, famous also for his artistic gifts) had crippled feet. The foot "is supposed to possess a magical generative power" that is specifically phallic.[12] Folklorist Stith Thompson observes that "Dwarfs...are sometimes spoken of as having their feet twisted backward...."[13] Indeed, this is how Hephaistos is pictured in certain Greek vase paintings, including an *amphoriskos*,[14] a cauldron,[15] a *hydria*,[16] and a fifth-century BCE *dinos* from Sicily.[17]

---

[10] Lotte Motz, *The Wise One of the Mountain: Form, Function and Significance of the Subterranean Smith: A Study in Folklore* (Göppingen: Kümmerle Verlag, 1983), 80.

[11] Ibid, 158.

[12] C. G. Jung, *Symbols of Transformation*, Collected Works of C. G. Jung, rev. R. F. C. Hull, Vol. 5, trans. H. G. Baynes (Princeton: Princeton University Press, 1967), 183.

[13] Stith Thompson, *The Folktale* (Berkeley: University of California Press, 1977), 248-49.

[14] Athens National Museum 664, Pl. 10.1, in Frank Brommer, *Hephaistos: Der Schmiedegott in der Antiken Kunst* (Mainz am Rhein: Verlag Philipp von Zabern, 1978).

[15] Rhodes 10.711, Pl. 11.1, in Frank Brommer, *Hephaistos: Der Schmiedegott in der*

In addition to the smiths' fundamental connection with the magical, poetic, and plastic arts, the "making" of the chthonic phallic cohort includes healing and foster-parenting. In various versions of this mytheme, Gaia (or Rheia) gives infant Zeus (or infant Poseidon) into the fostering care of the Telchines (or Kabeiroi), showing the nurturing aspect of the connection of these chthonic phallic divinities to the Great Goddesses.[18] It is from the "mother of the gods" that they receive their knowledge of the blacksmith's art.[19]

A key quality of the smith gods is the containment of opposites and ambivalences. The properties of iron include its red hotness and malleability, but it also very quickly becomes "white, hard, cold and unyielding."[20] The blacksmith's irreducibly opposite qualities include both "fiery" and "cool."[21] Ogun, a blacksmith deity from West Africa (whose cult remains alive in Nigeria, as well as undergoing transformation into an important factor in African Diaspora religions) is the patron of the hunter, the warrior, and the farmer. All three share in common their dependence upon iron implements, even though their aims and interests may come into conflict and they may regard each other ambivalently. Like his brother-smith, Hephaistos, Ogun, god of metalworking and by extension all technology, is a civilizer, who forges the scepter that is the symbol of royal legitimacy and thus of ordered society and the rule of law. He is also the god of chaos and conflict (think of the 'clash of iron' in battle), and his adherents acknowledge and revere his two-sided nature.

Like their brother and sister shaman gods, the blacksmith-artist gods function as mythic containers of ambivalent and seemingly irreconcilable qualities and attitudes: masculinity and femininity, hot

---

*Antiken Kunst* (Mainz am Rhein: Verlag Philipp von Zabern, 1978).

[16] Vienna M.218, Pl. 11.1, in Brommer.

[17] Würzburg H 5352, Perseus Vase Catalog, Perseus Digital Library Project, ed. Gregory R. Crane, 2004, Tufts University, accessed November 5, 2004, http://www.perseus.tufts.edu/.

[18] Gantz, 149.

[19] Jung, *Symbols*, 127.

[20] Henry John Drewel, "Art or Accident: Yoruba Body Artists and Their Deity Ogun," in *Africa's Ogun: Old World and New*, 235-264, ed. Sandra T. Barnes (Bloomington: Indiana University Press, 1997), 240.

[21] Adeboye Babalola, "A Portrait of Ogun as Reflected in Ijala Chants," *Africa's Ogun: Old World and New*, ed. Sandra T. Barnes, (Bloomington: Indiana University Press, 1997), 147.

and cold, chaos and order, violence and nurturance, art and technology. That their imaginal connections to the Hephaistos myth have been marginalized or forgotten points to the fragmentation of the mythic image of the artful maker once represented by the blacksmith/artist/shaman.

## Looking Forward from Homer

The Renaissance and Baroque Hephaistos appears with the revalorization of classical art and literature following the medieval era. The representations of Hephaistos in sixteenth- through eighteenth-century art—where he is more commonly identified as Vulcan, the Roman name for the originally Greek god—present a somewhat different set of preoccupations with the god's story than earlier imaginings. In later centuries (as also already in the Middle Ages), the classical gods were used as exemplars of the virtues (and sometimes the vices). Images of the gods were also adapted into alchemical texts. With the reappropriation of classical images in the Renaissance and Baroque periods, artists represented scenes from Homer and Ovid. Vulcan's iconography in artistic representations of the Olympian pantheon usually includes the blacksmith's hammer and tongs (and he sometimes wears his smith's apron amid the naked gods). Piero di Cosimo's *The Finding of Vulcan on Lemnos* (1495-1505) depicts the god as an adolescent discovered by Lemnian women after his fall from Olympus.

A favorite subject of artists of the Baroque period is the tale of the adultery of Aphrodite and Ares and Hephaistos' revenge, mentioned above, that Odysseus hears sung by the bard Demodocus at the court of Alcinous. Velázquez painted a version, *The Forge of Vulcan* (1630), showing the arrival of the all-seeing sun-god Helios, who has come to the forge of Hephaistos to inform him of the infidelity he has witnessed.[22] Joachim Wtewael's *Mars and Venus Discovered by the Gods* (1603-04) depicts the moment when the invisible net has dropped, the lovers are trapped luridly in flagrante delicto, and the Olympian gods have been assembled as witnesses—the fleshy physicality of the amorous pair and the laughing, naked gods is

[22] Diego Rodríguez de Silva y Velázquez, *Vulcan's Forge*, 1630, oil on canvas, 223 x 290 cm., Madrid, Museo del Prado.

vividly celebrated, presumably as a moral lesson on the risks of infidelity.[23] Several painters, including Francesco Solimena (*Venus at the Forge of Vulcan*, 1704), depicted Vulcan presenting glorious armor to Venus (Aphrodite).[24] (In the *Aeneid*, Venus secures armor for her son Aeneas, in Virgil's appropriation of the parallel scene in Homer's *Iliad* in which Hephaistos makes the shield Thetis will present to her son Achilles). A related subject, twice painted by François Boucher (*Venus Demanding Arms from Vulcan for Aeneas*, 1732,[25] and *The Visit of Venus to Vulcan*,[26] 1754) simply shows Venus and Vulcan together in the forge, presumably in the moment that Venus is making her request of Vulcan (either verbally, or by means of her seductive beauty). This scene has been described as an allegory of the triumph of love over power.

John Milton summons up Hephaistos, under another Roman name for the god, Mulciber, in *Paradise Lost* (1667). There Milton makes use of Homer's account of Hephaistos' wounding fall when he is thrown from Olympus by rageful Zeus, making this motif an echo of the fall of Lucifer from heaven. Mulciber, once the renowned architect of Heaven before he intrigued against Zeus, is now the architect of Lucifer's brazenly magnificent Hell.[27] In Milton's theme of the artist thrown from heaven and damned can be seen the faint outline of one thread issuing from the Hephaistos mythos that has stretched from the distant image of the smith-god and the chthonic phallic cohort—all creator gods associated with magic and mystery—into that part of the contemporary imagination that regards the artist with vestiges of ancient fear and suspicion, as a possessor of quasi-magical skills that carry with them a societal stigma of difference, defiance, and deviance.

---

[23] Joachim Anthonisz Wtewael, *Mars and Venus Surprised by Vulcan*, 1604-1608, oil on copper, 20.3 x 15.5 cm., Los Angeles, Getty Center.

[24] Francesco Solimena, *Venus at the Forge of Vulcan*, 1704, oil on canvas, 205.4 × 153.7 cm., Los Angeles, Getty Center.

[25] François Boucher, *Venus Demanding Arms of Vulcan for Aeneas*, 1732, oil on canvas, 252 x 175 cm., Paris, Musée du Louvre.

[26] François Boucher, *The Visit of Venus to Vulcan*, 1754, London: Wallace Collection, ArtRussia, accessed 11. December, 2004, http://www.artrussia.ru/en/picture_rarity/246.

[27] John Milton, *Paradise Lost*, 1667, I:730-751, Renascence Editions, 1992, University of Oregon, accessed June 9, 2003, http://darkwing.uoregon.edu/~rbear/lost/lost.html.

## The Wounded Artist

A development in the history of ideas is the philological and scientific interest in myth and mythogenesis as a feature of an earlier stage in human existence. F. Max Müller, the nineteenth-century mythographer, and, slightly later, Lucien Lévy-Bruhl, the early twentieth-century philosopher, psychologist, and ethnologist, used the term *mythopoesis* to characterize what they supposed to be a feature of primitive human thought, "wherein myth and metaphor, rather than the supposedly later-developed science and *logos*, became dominant."[28] By 1913, both Sigmund Freud and C. G. Jung recognized myth and symbol as a pathway to the contents of the unconscious. (That they differed in the specific nature of this pathway may have formed part of the ground on which they famously broke their previous professional and personal relationship.)

Freud was fascinated by the myth of the artist as a species of sacred monster, certain of whose creative energies and/or creative inhibitions could be seen to be lodged in pathology.[29] Freud himself engaged in the exercise of artist psychobiography, notably of Leonardo da Vinci, whom he diagnosed as an obsessional neurotic and narcissist due to a precocious eroticization in his early and isolate relationship with a doting mother. In this first foray into the arena of the psychoanalytic biography, Freud defended himself against the potential claim that he was dragging genius in the mud by identifying the source of both Leonardo's creative expression and its inhibition in a precocious infantile sexuality.[30] At the same time, Freud acknowledged that "I am a scientist by necessity, and not by vocation. I am really by nature an artist." The proof of this statement was, for Freud, the observation that "in all countries into which psychoanalysis has penetrated it has been better understood and applied by writers and artists than by doctors."[31] While Freud seemed

---

[28] William G. Doty, *Mythography: The Study of Myths and Rituals*, 2nd. ed. (Tuscaloosa: University of Alabama Press, 2000), 20.

[29] Lionel Trilling, "Art and Neurosis," *Art and Psychoanalysis*, ed. William Phillips (New York: Meridian, 1963), 504.

[30] Sigmund Freud, *Leonardo da Vinci and a Memory of His Childhood*, trans. Alan Tyson (New York: W.W. Norton, 1964), 13.

[31] Sigmund Freud, qtd. in James Hillman, *Healing Fiction* (Woodstock: Spring, 1983), 3.

in fact to hold artists together with their magnificent sublimations in a sort of awe-ful reverence, Jung, while in fact busily creating art—stoneworks and sculptures, daily mandalas, the voluminous Red and Black Books—in pursuit of the mysteries of his own myth, emphatically denied the artist-image. He conflated it, negatively, with the voice of his personal anima—the contrasexual, thus feminine, voice of his unconscious—which troubled him with the taunt that he was making "art" and not "science" during the time of his personal sojourn and cotemporaneous self-analysis through a period of terrifying waking visions and confidence-shattering dreams.[32]

Henri Ellenberger imaginally identified both Freud's and Jung's creative breakthroughs with the shaman-healer's "creative illness."[33] Neither mentioned Hephaistos in connection with their thinking and writing about art, but the contributions of both have had profound ramifications in regard to how contemporary society views art as an expression of the unconscious, and artists as sometimes fruitfully, sometimes threateningly, sometimes pathologically close to its awesome contents. For some depth psychologists, Hephaistos, the crippled god of craft, has specifically come to signify the wounded psyche of the artist. Indeed, the wounded artist is taken as a visible symptom of society's woundedness, a sort of canary in the coal mine of malaise of soul in an age cut off from meaning. Depth psychologist Murray Stein has written about Hephaistos as the archetype representing the pathology of the mother-wounded psyche of the introverted, angry, sexually compromised male artist. Child psychotherapist Valerie Sinason invokes the name of Hephaistos in her analysis of the deep-seated fear in the minds of many parents of congenitally impaired children that some flaw in their sexual and procreative connection (like the mythic anger between Zeus and Hera) caused the damage to their offspring. In the same way that the illegitimate have had to carry social and cultural fantasies about wild, unlawful sexuality, so the congenitally different carry, stamped on their bodies, the mark of what is feared by society as unnatural and

---

[32] C. G. Jung, *Memories, Dreams, Reflections*, ed. A. Jaffé (New York: Vintage 1989), 187.

[33] Henri Ellenberger, *The Discovery of the Unconscious* (New York: Basic Books, 1970), 889.

destructive sexuality. They also carry a reciprocal anger toward the whole-bodied.[34]

More recently still, it has been suggested that the more troubling as well as the more profoundly creative aspects of certain artists' personalities and accomplishments have an organic origin in a type of epilepsy, originally known as Geschwind's Syndrome after the neurobiologist who first described it, and now known as Temporal Lobe Syndrome (TLS). Epilepsy was traditionally thought of as a type of divinely inspired condition, productive of numinous visions and creative madness. It has more recently been classified as a condition that is sometimes heritable (thus a "deformity"), sometimes caused by injury or other factors, that interferes with the electrical activity of the brain. It is now postulated that Vincent Van Gogh, Fyodor Dostoievsky, and even Lewis Carroll may have suffered— and posterity benefited—from TLS[35].

This mechanistic perspective on the nature of mind has links with another Hephaistean thread discernable in contemporary society, that deriving from ancient myths of the "magical" aspects of technology, its origins, and its effects, both positive and negative. Some of the suspicion of the artist is a remnant of the ancient creator-artist gods' ability to make uncanny imitations of life, and the ancient blacksmith-sorcerer gods' capability of cursing life.

Monstrous Technology

In the United States, members of the post-WWII "baby boom" and earlier generations are most likely to associate the god's name with heavy-industrial products or combustive processes expressive of Vulcan's volcanic fire, for example "Vulcanized" rubber tires, Vulcan Motor Oil, or the statue that since the 1907 World Exposition has towered over Birmingham, Alabama as emblem of the city's metal-forging industry (and inspired Birmingham's local "Vulcan" popular-music record label). Today, Vulcanized tires and Vulcan Motor oil are no longer found in the marketplace, and in the high-

---

[34] Valerie Sinason, "Challenged Bodies, Wounded Body Images: Richard III and Hephaestus," in *Splintered Reflections: Images of the Body in Trauma*, eds. Jean Goodwin and Reina Attias (New York: Basic Books, 1999), 183.

[35] Eve LaPlante, *Seized: Temporal Lobe Epilepsy as Medical, Historical, and Artistic Phenomenon*, (New York: Harper, 1993), 1-10.

priced labor economies of North America and Western Europe, heavy industry is disappearing from the scene. However, a casual tour of the Internet reveals that Hephaistos (indeed, on occasion under his Greek name), often appears amid the panoply of figures associated with computer-generated and Internet-resident fantasy role-playing games, transformed into innumerable, generally robotic guises that tend to be heavily armored and are often flame-throwing. An example is game-developer Nvidia Corporation's "Vulcan" promotional animation designed to highlight the capability of its software to render "realistic fire, smoke, and glow using volumetric texturing and render-to-texture techniques."[36]

Though large parts of the Hephaistean mythos have seemingly been forgotten or dispersed, the psychic interrelationship between technology, magic and creativity—as well as the shadow of this conjunction—can readily be discerned in ideas and artistic productions and indeed in daily headlines which never name the god, but are nevertheless inspired by him.

The Hephaistean theme of technology naturally involves images and analogues of forging and metalsmithing, and specifically of arms manufacture (hence the association with heavy industry and military materiél). It also involves, in contemporary analogues, robotics and bioengineering. Some of Hephaistos' mythic creations have humanoid forms. Pandora, the archetypal woman, created by Hephaistos from clay at Zeus's orders, is meant to bedevil mankind, who otherwise might pretend to the status of gods after having received the knack of controlling the fire stolen by Prometheus from Hephaistos' forge. The mechanical forge assistants Hephaistos fashions out of gold are said to be both beautiful and intelligent.[37] Hephaistean mythemes are notably implicit in science fiction, dating from the beginning of the industrial age. The mytheme of humans arrogating to themselves the powers of gods is ubiquitous but the attempt to *fabricate* life from scratch, as it were, is specifically Hephaistean. Mary Shelley's 1818 *Frankenstein* is the tale of the hubristic attempt of a young physicist to create life, with dire consequences. The stories of E.T.A. Hoffmann, written in the first

---

[36] "Vulcan Demo," nZone, © 2003, 2004 NVIDIA® Corporation, accessed December 6, 2004, http://www.nzone.com/object/nzone_vulcandemo_home.html.
[37] Homer, *Iliad*, trans. Fagles, 18:488-93.

two decades of the nineteenth century, concern sinister scientists and uncanny humanoid dolls so lifelike that humans are induced to fall in love with them, a fatal mistake (just as Zeus intended when he ordered the creation of Pandora). Hoffmann's most famous tales were later recast into Jacques Offenbach's opera *Les Contes d'Hoffmann*, Leo Delibes' ballet *Coppelia*, and Tchaikovsky's *Nutcracker Ballet*. The nineteenth-century fascination with automata has persisted into the twentieth and twenty-first centuries, witnessed by the continuing popularity of the opera and ballets mentioned, and of science fiction. Possibly best exemplified in popular cinema, the fascination with the fictional robot (*Metropolis, The Day the Earth Stood Still, AI*), the cyborg (*Terminator, RoboCop, Star Trek*), the bioengineered humanoid (*Frankenstein, Blade Runner*) and the clone (*Star Wars: Attack of the Clones*, the daily newspaper, and recent and ongoing congressional hearings on stem cell research) is a fascination that has deeply Hephaistean roots.

Terry Gilliam's 1989 film, *The Adventures of Baron Munchausen*, features a visit to Vulcan's forge in Mount Etna where the Baron inadvertently interrupts a labor dispute between Vulcan and his Cyclops forge workers, who meanwhile are fashioning a nuclear ballistic missile. Sten Nadolny's 1994 novel *The God of Impertinence* features a Hephaistos imaginatively metamorphosed into the gruff, cigar-chomping Manager, Lord of the present day Universe. He is the only one of the Olympian gods left minding the store in the wake of Zeus's retirement to play golf in Iowa and amid the progressive ennui that causes the other immortals to absent themselves (except for Hermes, whom Hephaistos has immobilized in a rock face for 2,791 years). Hidden behind the scenes, Hephaistos the Manager unilaterally manipulates the global economy toward a holocaust that would put an end to the gods' terminal boredom (and to all life on earth in the bargain).

These are examples of Hephaistos as possessor of hidden, instrumental knowledge, maker of war matériel and, perhaps above all, servant to the Olympian power of his father Zeus.

## Two Mythopoetic Themes

Two related themes emerge as aspects of recent Hephaistean mythopoeticizing. One concerns the artist, who in the popular imagination is viewed ambivalently, on one hand as a maker endowed with a god-given gift of creativity that relatively few people possess. On the other hand, and often at the same time, the artist is viewed with suspicion and distrust: as an outsider, a potentially pathologically disruptive character whose removal from society may at times be seen as a positive or necessary thing, or as a consequence of the artist's own fatal nonconformity.

The second theme involves contemporary images of technology and its applications. The Hephaistean archetype is clearly present in the guise of the corporate manager (as exemplified in Nadolny's fiction), who increasingly signifies the impersonal, intertwined forces of capital, political influence (often clandestine), and advanced technology. As developer of Vulcan's prototype nuclear ballistic missile, the god is also imaginally implicated as complicit in the creation of the ultimate technological fire—'weapons of mass destruction.'

Rich as they are, there is something missing in all of these examples of Hephaistean mythopoesis. It is as if the progress of the god, at least from the time of the Renaissance until today, has proceeded without true consciousness of the divine quality of imagination itself, specifically the divine imagination that produced the fantastic shield of Achilles and inspired Homer's breathtaking description of its amazing conception. What is more, it seems as if the further in time the remythologizing is separated from the ancients, the less magical and fleet of imagination Hephaistos appears, and the more grimy. He has become pathologized into an image of creative and relational blockage, through anger or repression, or alternatively, into the willing and willful purveyor of at best prosaic, at worst deathly materiél. No longer is he seen to serve the imaginative purposes of the maker, whose engagement with matter issues in beautiful, useful, appropriate and enlivening works of art and industry. The image of the divine blacksmith-artist has been sooted into unrecognizability.

In the wake of the final dismemberment of the myth of the divine smith, in a time when Hephaistos is seen as representative of destructive, rather than creative, instrumentality, I see a society in which most people deny the divinity of human creativity, attribute it to others, leave it to the experts, and fear and vilify the artists who are its practitioners without really knowing why. In the final fragmentation of an ancient creation myth, divinity is denied, humanity-as-divinity is denied, and so is the responsibility of individual humans to assist in the ongoing creation. Were the myths of contemporary Western culture to view the world as a work of divine artistry carried on by the divinized makers in an ongoing and acknowledged mythopoesis, owning its part in the ongoing co-creation, would we live in a world that continues to be broken, dying, violent, gasping?

The Hephaistean mythos contains a key to the origin of images of and attitudes toward art and technology. The ill-effects of a broken ethics are the result of the ever-increasing separation of the fruitful, erotic fantasies of art from the practical applications of technology. Though the alienation of technology from art (and, I will suggest, thereby from ethics) is by no means a recent development historically, it was nevertheless not always so. Retrieving the full spectrum of the Hephaistean mythos has many implications. What can be gained? The image of the blacksmith god is an image of wholeness. It acknowledges and contains seemingly irreconcilable opposites and ambiguities. The recovery of this archetype in its full dimensions may provide an image for the reuniting of art and technology. Ultimately, it has the potential for providing an image that could provide an exemplar for the holistic attitude toward ethical technology that will be necessary to sustaining life in our profoundly interdependent cosmos.

What can heal the separation? Perhaps, and perhaps only, the deliberate recovery and internalization of mythic and mythopoetic images of humanity's discredited capability to contain in its collective psyche more than one view at a time; its ability to contain fertile ambivalence: hot *and* cold, male *and* female, human *and* nonhuman, life *and* death, art *and* technology.

## Literature Review

This dissertation aims to reconstruct a comprehensive Hephaistean mythos from the many strands outlined above. By tracking the myth from the time of Homer into the present day, I will show that the myth of Hephaistos is a living one. In fact, it is a potent archetype, even though the god is rarely mentioned by name in the myriad cultural productions he inhabits and the attitudes he inspires.

The work of many writers has suggested there is more to be seen in the Hephaistean archetype than meets the eye in the still-popular works of Homer and Ovid. Marie Delcourt has thoroughly explored the surviving classical Hephaistean texts and images and has convincingly advanced a thesis she terms "Hephaistos the Magician." Murray Stein has contributed a depth-psychological portrait of Hephaistos as archetype in the soul of the mother-wounded male artist. Jack Denslow has shown the wondrous shield made by Hephaistos to be an emblem of psychic individuation and wholeness in the Jungian sense. My intention in this dissertation is not to duplicate the very useful work already done by these and many other scholars. My aim is to use this work to reweave the fragments of myth into a whole and multilayered fabric of story, in which the face of a living and powerful archetype may be discerned: Hephaistos, the god of fire and forge, art and technology and above of all the effective power of the imagination manifested in matter—in short, the mythos of making.

Although there is little literature specifically devoted to Hephaistos to review per se, I will engage the work of many authors—from the ancient sources of Hephaistean and other myths to commentators on topics that will be shown to be relevant to reconstructing Hephaistos' myth.

Chapter 2 will present and discuss the work of several authors, for the purpose of introducing key Greek terms with relevance for the argument I will develop in regard to the lineaments of the Hephaistean archetype.

In tracing the outlines of the archetype, I will of necessity be tracking the progress of the mythopoesis of the Hephaistos image complex. In looking backward from Homer to gather scattered fragments of myth and tradition, and in making lateral connections to mythic traditions which are related by theme but not necessarily by traceable historical connections, I will of necessity engage in interpretation, which is itself an exercise of imagination. Thus, the second concern of the present work will be not only to define the term, but to examine the philosophical and depth psychological purposes of mythopoesis as itself a signally—and in certain ways uniquely—Hephaistean activity.

An examination of the Hephaistean image from the historical starting point of Homer's use of the god's myth in the *Iliad* and *Odyssey* reveals an archetype split into two disconnected parts, the artist and technologist. It also reveals something that has been left out, or left to be expressed in the shadow of the archetype. What is left out of the conscious mythopoetic engagement with the Hephaistean archetype since Homer is the powerful, unpredictable element of the magical imagination.

The mythopoetic themes of *wounded artist* and *monstrous technology* have evolved not only from the Homeric Hephaistos but also to some degree from inconsistencies between aspects of the Hephaistean myth presented in Homer and those presented in Hesiod and other Greek and subsequent sources. One of the aims of this dissertation is to recover and examine aspects of the Hephaistean mythos that are not necessarily part of the Homeric Hephaistos but are a part of the surviving corpus of Greek and Roman, and where possible, earlier texts and visual depictions. Examination of the divergent accounts of Hephaistos' parentage, and the introduction of non-Greek material on the ubiquitous myths of the blacksmith-creator divinities, opens a number of avenues into the reconstruction of a more comprehensive Hephaistean mythos.

Although the term mythopoesis combines two Greek words: *mythos* and *poiēsis* mythopoesis is not a Greek term. It is of modern coinage, meaning "myth-making." In the last volume of his *The Masks of God*, Joseph Campbell writes about what he terms "Creative Mythology," and provides what I take to be a definition of mythopoesis. Campbell identifies four interactive functions of myth, two of which are especially important for this study. For Campbell, one of myth's functions is to enforce a moral order within each successive coherent social group, within the context of its unique geographical and historical place. Moreover, he believes, "The rise and fall of civilizations in the long, broad course of history can be seen to have been largely a function of the integrity and cogency of their supporting canons of myth." For Campbell, a mythic canon persists as long as it has the power to inspire individual members of a group to experience a harmonious connection with the social order and their place in the universe.

At the same time, the canonical myth is subject to change. This happens when the canonical myth ceases to hold meaning for at least some members of the group or produces "deviant" effects, giving rise to both a sense of dissociation and an urge toward a renewed quest for "meaning." Campbell points toward historical times of dissolution as productive of new myths. In other words, one of the functions of myth is to generate new myths. For example, in Christian Europe in the twelfth century, "beliefs no longer universally held were universally enforced." Many people lost faith in Scripture, giving rise to what Campbell labels the "Waste Land" theme in the imagery of the Grail legend: a theme of spiritual death, of wandering, and "waiting, waiting." Campbell asserts that Jean-Jacques Rousseau's 1749 work *Discours sur les arts et sciences* marks a further dissolution, this time a loss of faith in reason. Society came to be seen as a corrupting influence, and new myths appeared, of the "noble savage," the "natural man."[38] These myths expressed a desire to return to an imagined state of nature—a concept that was to be further articulated in the speculations of F. Max Müller, in the century following Rousseau, on the nature of the primitive mind and its connection with the numinous.[39] This epoch of dissolution, for Campbell, has so far lasted into the present day. The mythic theme of distress in the wake of loss of faith that inspired *Parsifal* is the same that impelled T.S. Eliot's creation of his 1922 work *The Waste Land.*

When change comes, it originates with individuals. Campbell's "creative mythology" is not a product of the authoritative voice of theology, but arises from the "insights, sentiments, thought, and vision of an adequate individual, loyal to his own experience of value."[40] This sincere individual voice provides a corrective to empty, left-over shells of forms no longer animated by living and breathing spirits. The literary critic Harry Slochower offers a definition similar to Campbell's. For Slochower, mythopoesis, whether performed by Homer or by Joyce, is the transformational re-creation of ancient

---

[38] Joseph Campbell, *The Masks of God: Creative Mythology*, (New York: Penguin, 1968), 5.

[39] Doty, *Mythography*, 11.

[40] Campbell, 6-7.

stories into works with coherent symbolic meaning. Slochower appears to reflect on the political and cultural mythology of his own time (his book *Mythopoesis* was published in 1970, two years after the publication of Campbell's The Masks of God: Creative Mythology). Standing at the fluid margins of the vivid political rallying ground of popular art and music created in protest against the advance of war in Vietnam, and the growing cultural fantasy of raucous revolution at home in the United States, Slochower asserts that mythopoetic works such as the Sophoclean tragedies, the Book of Job, The Divine Comedy, Don Quixote, Moby Dick, The Magic Mountain, and others,

> arose when the literal account of the legend could no longer be accepted. They arose in periods of crisis, of cultural transition, when faith in the authoritative structure was waning. It is at this juncture that our great prophets and artists would redeem the values of the past and present in their *symbolic* form, transposing their historic transitoriness into permanent promises.[41]

For Campbell and Slochower, the individual myth-maker—read artist, novelist, poet—is the creator of new mythic meanings. More recently, Chris Baldick challenges this "Romantic" notion, pointing out that essentially new myths created in modern times—Faust, Don Quixote, Robinson Crusoe, Frankenstein, Jekyll and Hyde, Dracula— are expressions of a larger cultural impulse and not only of the visions of the individual creator. Baldick traces the creation of the Frankenstein "myth" half a century backward from its literary creation by Mary Shelley to the metaphors that arose out of the social and political upheavals that resulted in and followed the French Revolution.[42] These metaphors, condemning the human "monstrousness," both of the part of revolutionaries and of the abuses of the system they meant to overturn, were part of the culture at large when Shelley took up her pen to write her Hephaistean story of a monster assembled from human parts and brought to life by an

---

[41] Harry Slochower, *Mythopoesis: Mythic Patterns in the Literary Classics* (Detroit: Wayne State University Press, 1970), 15.
[42] Chris Baldick, *In Frankenstein's Shadow: Myth, Monstrosity, and Nineteenth-century Writing* (Oxford: Clarendon, 1987), 16-17.

earnest student of natural science who came to despise and fear his creation as soon as he had accomplished it.

Social conditions generate new myths that "live" us in the sense of Jung's great question, in *Memories, Dreams, Reflections*, What myth is living me?[43] Each successive overarching mythos functions dialectically, both prescribing social behaviors and calling for correctives. These correctives are articulated by individuals and over time in given contexts result in new myths.

I will also borrow from Mircea Eliade's exhortation that we study other cultures' experiences of the sacred to understand *ourselves* better and to open new pathways for our own search for the sacred. "It is," he says, "the *personal* experience of this unique hermeneutics that is creative." This can be read as a challenge to religious scholars to pass beyond values of "obsolete reductionism" and "pure erudition" and directly participate in the dialogue with the data of religious experience, making "an effort to understand them *on their own plane of reference.*"[44] According to Eliade, the artist and the depth psychologist have preceded the scholar into the inevitable future of the scholars' hitherto carefully delimited field.

The definition of mythopoesis I will use here combines those put forward by Campbell, Baldick, and Eliade: mythopoesis, literally the making of myth, involves a creative act which expresses itself within the context of specific historical and cultural contexts. While rooted in the present and past, it also looks forward. As William Doty recognizes, both the backward looking and forward looking aspects of "rituals, symbolic images and myths establish conservative benchmarks, but at the same time they anticipate forms of the future as they determine and shape ideals and goals for both individual and society."[45]

It is not enough to tell a story; the story, in order to live, must evoke a passionate response in its audience. While myth must be bound to culture through a dance of old and new, there is a difference between myth and mere cultural commodity. The *polis*—the

[43] Jung, *Memories,* 171.

[44] Mircea Eliade, "A New Humanism," in *The Insider/Outsider Problem in the Study of Religion: A Reader*, ed. Russell T. McCutcheon, (London and New York: Cassell, 1999), 96-97, italics mine.

[45] William G. Doty, "What Mythopoetic Means," *Mythosphere*, 2.2 (2000): 258.

community as political body—together with the individual creator, has a stake in the transmission of new myths. *Vide* Hephaistos, both old and new, an embodiment of both very old and very new and compelling truths.

## Organization of the Study

In the overall organization of this study, I will make use of David L. Miller's schema for re-evaluating a mythic theme through a forward-and-backward examination of its mythopoesis from a significant historical starting point. This schema supplies a useful method for re-assessing possible meanings of the Hephaistean mythos that have been lost or distorted, as well as organizing what has remained in the foreground.[46] In this process, Miller identifies both the central idea and image representative of the archetype's manifestation. In this case, the idea is the mythic idea of Making; the rich and complex, undoubtedly erotic dance of imagination and matter, both divine and human. The image is of course that of the fire-blacksmith-artist god Hephaistos. It must be stressed that this idea can and indeed must be representable by other images besides only Hephaistos. Archetypes do not exist as psychic islands: they are always interactive and interconnected. At the same time, it will be shown that the multivalent characteristics of the Hephaistean image are expressive of a unique and specific set of qualities and attitudes that bear on the nature of creative instrumentality and its expression through technology, whether the making be a bomb or a poem.

In Miller's schema, an idea and an image come together at a particular time in history that sets their paradigmatic interrelationship. Miller uses the example of the Christian idea of "Perfection" as expressed in the image of the Good Shepherd.[47] He next investigates the conjunction of image and idea at its source, in the case of the Good Shepherd example in its appearance in first century Christian art. It is useful, Miller demonstrates, to look forward in time to discern the mythopoetic development of the image, for example into the two strands of pastoral love poetry and

---

[46] David L. Miller, *Christs: Meditations on Archetypal Images in Christian Theology* (New York: Seabury, 1981), xix-xxi.
[47] Miller, *Christs*, 3-7.

the historical development of Christian theological notions and interpretations of perfection as the imitation of Christ the Shepherd. It is useful as well to look backward from the thematic pivot-point, again in the case of Miller's example of the Good Shepherd to the ancient purposes of flocks and animal fecundity, divinities manifest in animal forms, and their relation to human social needs as expressed in myth and ritual. For example, there are hints of ritual re-enactment of the copulation of Hermes the ram-god and the ewe-queen in ancient Samothrace (a Kabeiric, and thus, as will be shown, a crypto-Hephaistean cult center), a union which issued in the line of descent of its rulers.[48] Further, the shepherd, far from being a model of "good," was in the ancient Near East an emblem of the opposite: untrustworthy and often socially outcast. Accordingly, Miller discerns a split between two archetypes that were thrust together in an unnatural manner: the archetype of the sheep—who belongs with the ram, or the ram-god—and the archetype of the shepherd in the pastoral field who belongs with the shepherdess—where he should by nature and expectation be seducing the shepherdess and ignoring the sheep.[49] The split that Miller discerns points toward a shadow in the image of the Good Shepherd and a flaw in the associated idea of Perfection, revealed through an examination of the myth and mythopoesis of the image.[50]

Based generally on Miller's schema, the chapters of this dissertation will be arranged as follows:

Chapter 1 has introduced the myth of Hephaistos, the Greek god of art and technology. In Greek myth, as chiefly transmitted through Homer's *Iliad* and *Odyssey*, Hephaistos represents an ambiguous image, especially in regard to sexual and instrumental potency. Looking backward and sideways from Homer ancient traditions point to forgotten or fragmented aspects of the Hephaistean mythos: mystery and magical potency rooted in the association of the primordial feminine with the chthonic phallic. Two strands issuing from this ancient myth and discernable in contemporary culture, the

---

[48] Ibid, *Christs*, 29.
[49] Miller, *Christs*, 41-42.
[50] Ibid, *Christs*, 3-5.

"wounded artist" and "monstrous technology," have also been introduced, to be further explored in this dissertation.

Chapter 2 will introduce Greek terms basic to a renewed understanding of the Hephaistean archetype. In the case of each term, I will rely principally on the work of an author who has presented a book-length exploration of the term: *technê* (David Roochnik), *mêtis* (Marcel Detienne and Jean-Pierre Vernant), *poiēsis* (Lotte Motz), and *ekphrasis* (Andrew Sprague Becker). Another pair of Greek terms, *mythos* and *logos*, will be shown to have had their meanings significantly altered between the time of Homer and Plato (Bruce Lincoln). These alterations will serve to elucidate both the fact and the nature of the fragmentation of Hephaistos' mythos already in ancient times. Finally, I will introduce a brief review of the nature of Plato's and Aristotle's thought on the imagination. That the poetic imagination—which is to say the imagination of the maker—came to be discredited by the fourth century BCE is a critical factor in reconstructing the lost or overlooked aspects of Hephaistos' mythos.

Chapter 3 will examine the most significant surviving Hephaistean mythemes as preserved in relatively brief mentions in Hesiod's *Theogony* and, much more extensively, in Homer's *Iliad* and *Odyssey*. Smaller fragments are scattered among the writings of the authors of the so-called Homeric Hymns, Apollodorus, Pausanias, Pindar, and others. Additionally, about 800 visual representations of the god from ancient Greece and Rome are known (as catalogued by Frank Brommer), in stelae, marble sculptures, bronzes, as well as 280 vase paintings which themselves provide sometimes puzzling additions to Hephaistos' fragmentary mythos. These will be explored for key Hephaistean themes with reference to interpretations put forward by classicists and theorists including Frank Brommer, Walter Burkert, Marie Delcourt, Marcel Detienne, Jane Ellen Harrison, Nicole Loraux, Sarah Morris, Jean-Pierre Vernant and Pierre Vidal-Naquet.

Chapter 4 will look backward from the Homeric version to show how Hephaistos is representative of the primordial blacksmith/shaman/artist complex of maker/creator deities. Further, Hephaistos will be shown to be representative of what I will term the primordial "chthonic phallic cohort" of deities (with reference to Erich Neumann's use of the term "chthonic phallic"), whose mythic functions are represented in Greek and other mythologies and folklores and include magic and healing, and whose gifts derive from Mother Earth. Together, these deities form a complex in Jung's sense of the term, which exert a sort of hidden but powerful gravitational pull on the human psyche. These energies, once examined, must be incorporated into a reconstructed Hephaistean mythos.

In Chapter 5, looking forward from the historical starting point of the Hephaistos myth as presented in Homer's *Iliad* and *Odyssey*, I will trace two significant mythopoetic strands of the Hephaistean image that reflect ambivalences present in Western thought since the time of Homer. My main emphasis will be in examining the image of the "wounded"/pathologized artist developed in the depth psychological tradition of the twentieth century and exemplified in the writings of Freud, Jung, and Jungian analyst Murray Stein, who has written on what he terms "The Hephaistos Problem." I will also show that mythic Hephaistos survives in contemporary attitudes toward the exhilarating and frightening boundary-stretching of the newest technologies, and that it is specifically implicated in the images of cybernetics, biotechnology and the "military-industrial complex." Then, I will examine archetypal psychology, specifically querying James Hillman and Mary Watkins for insights into the psyche's mythopoetic capacities. Finally, I will present examples of contemporary art and technology that are representative of the reconstructed mythopoetic image of Hephaistos, the much maligned and wounded archetype whose image has powerful significance for a contemporary re-imagining of the inextricably interconnected relationship of art, technology, and ethics, and for full acknowledgement of the awesome creative power of the human imagination.

The Greek word *mythos* signifies "mouth" i.e., "of the mouth." Chapter 6 will present a reconstruction of the myth of Hephaistos, the ambivalent, imaginative Greek god of art and technology, in the first person—from the mouth of the god.

CHAPTER 2

=====OO=====

# Review of Greek Terms

THE PURPOSE of this chapter is to examine a set of terms in order to lay the groundwork for a recovery of the Hephaistean archetype. Homer frequently attributes to Hephaistos the epithet *klutotechnês*, meaning "famous for his art," or "renowned artist" and containing the root *technê*, from which is derived the word "technology." An epithet also applied to Hephaistos is *klutomêtis*, "famous for skill," which contains the root *mêtis*, meaning, roughly, "cunning" or "intuitive intelligence." These and other variants on the terms *technê* and *mêtis* appear repeatedly in connection with Hephaistos. This pairing is significant, for as will be seen, these terms are divergent in meaning, *technê* referring to what is systematic and repeatable, as in a set of technical skills, and *mêtis* referring more to a way of thinking that is oblique and unpredictable in nature. Some of the key questions that an examination of these terms will help to answer are the following: How is Hephaistos to be understood as an archetypal representative of technology? What is the "field" of technology—both in the determinative, left-brained, linear sense of a *technê* and in the right-brained, shape-shifting, nonlinear sense of *mêtis*—as represented in the archetypal image of Hephaistos?

A third Greek term that will be discussed in this chapter is *poieô*, "to make," from which is derived the more familiar term *poiēsis*, representing the actions resulting in products to which Hephaistos applies his *technê* and *mêtis*. It will be seen that, far from being restricted to the specific activities of the forge, Hephaistean making as *poiēsis* embraces a broad conception of making, which includes the artistic and poetic, a word which derives from *poieô*. It can include the making of meaning, including religious and sacred meaning, which in turn reflects upon the significance of Hephaistean objects. A fourth term, *ekphrasis*, loosely meaning "lively description" as part of the art of the poet will be seen to provide insights into the combination of *technê* and *mêtis* in poetic making. The questions these terms will help to answer include: What are the unique qualities of Hephaistean making? How does the Hephaistean archetype express itself in poetics, and why is this significant?

Next, I will explore the meaning(s) of mythology by examining its roots-words, *mythos* and *logos*. These are terms that will be shown, like *technê*, to have been subject to fundamental changes in meaning since the time of Homer. In fact, the terms *mythos* and *logos* almost exchanged their meanings. *Mythos* was once understood to be a divinely inspired and thus authoritative mode of speech. *Logos* formerly denoted the speech of women and weaklings whose only method to persuade was through "seductive wiles." The shift in meaning of these terms, such that *mythos* became discredited and suspect, and *logos* privileged, will be examined for its implications in regard to the discrediting of myth generally and its effect on ancient and modern attitudes toward myth as a way of knowing. Finally, I will examine views of Plato and Aristotle on the creative imagination.

## Technê

The word technology comes from *technikon*, meaning that which belongs to *technê*. *Technê* makes its literary appearance in the time of Homer, Hesiod (c. 850-950 BCE) and the later, unknown authors of the so-called Homeric Hymns. The *technê* of Hephaistos is celebrated in the Homeric Hymn to the god:

> Sing, you clear-voiced Muse, of Hephaistos renowned for craft
> [*klutomêtin*],
> Who with bright-eyed Athena taught splendid works to
> humans on earth —
> They had before been dwelling in caves on the
> mountains like beasts,
> But now, knowing works through Hephaistos renowned for his

skill [*klutotechnên*], with ease
Till the year brings its end they live in comfort within their
own homes.
Come now, be kindly Hephaistos; grant us prowess and
Wealth.[51]

Appearing in written form for the first time perhaps between the mid tenth to ninth centuries BCE, *technê* is a word of much older derivation whose meaning had already altered considerably by that time. The word likely derives from the Indo-European root *tek*, which is first associated with working in wood and which means "to fit together the woodwork of a woven house." This house construction is thought to have been a community act, accomplished by utilizing skills originally held within the collective. Later on, the term came to denote the knowledge possessed by an individual whose specialized function within the community was to construct things from wood, the *tektōn* or "woodworker."[52]

Lewis Mumford observes that, "The building of Neolithic villages on wooden piles over the waters of lakes was one of the surest witnesses to the advance of civilization: wood delivered man from servitude to the cave and to the cold earth itself." Added to the malleability and strength of wood, its combustibility "was more important and more favorable to human development than the fire-resistance of other materials." The toughness, tensile strength, and natural adaptability of different species of wood readily provided the *tektōn* with more primary technical possibilities than metallurgy would be able to provide until it had long evolved; indeed, the dominance of key technologies based on the use of wood, for example shipbuilding, persisted into the nineteenth century. Mumford notes that "Wood gave man his preparatory training in the technics of both stone and metal: small wonder that he was faithful to it when he began to translate his wooden temples into stone."[53]

Yet, possibly because wood is impermanent, unlike stone and metal, many cultures have imagined themselves in terms of metal. R. J. Forbes reports that "The idea of dividing the history of the world in

---

[51] "Hymn to Hephaistos," trans. Crudden, 84-85.
[52] David Roochnik, *Of Art and Wisdom: Plato's Understanding of Techne* (University Park: Pennsylvania State University Press, 1996), 19.
[53] Mumford, 78-79.

different periods named after metals is probably of Iranian origin" and it "recurs in Buddhist doctrines."[54] According to Hesiod, the immortals made a "golden race of mortal men" who lived without care or toil until death overtook them like sleep. A second race was created of silver, "a much worse one." Immature and foolish, the people of this race were violent and impious, refusing to worship the gods or perform sacrifice, for which reason Zeus "buried" them, and they too passed from existence. Zeus then made a third race, of bronze, "dreadful and mighty and bent on the harsh deeds of war and violence," who killed each other off. They were succeeded by the divine race of heroes, who died in battle, "some fighting over the flocks of Oidipous" at Thebes and some under the walls of Troy. The present race is a race of iron, among whom the just will not prosper. Humanity now toils under the load of envy—and Hesiod wishes he were born earlier or later.[55]

Mumford asserts that "The rational conquest of the environment by means of machines is fundamentally the work of the woodman," by virtue of the civilizing technological developments wood's manipulation made possible.[56] Yet, as Forbes points out, though the products of metallurgy were slow, superseding products made of stone, because metals are more malleable and transportable while being as durable, "metallurgy accompanies the rise of urban civilization and the formation of the first empires in history." Metals enabled long-distance trade and the formation of social classes based on the establishment of wealth. Coinage of precious metals (i.e., the assurance of a constant value by dint of standards for purity and weight of metals and the use of stampable hallmarks of origin) served a role in the "accumulation of wealth in the hands of the few," and, "indeed, much of ancient history could be rewritten as a struggle for the domination of quarries and ore-deposits or metal-supplies."[57]

Bronze provided weapons, tools, and "luxuries for a small wealthy class." The discovery of smelting iron opened up far greater technical and economic possibilities. Copper and tin, the essential constituents

---

[54] R. J. Forbes, *Studies in Ancient Technology*, Vol. 8 (Leiden: E. J. Brill, 1964), 1.

[55] Hesiod, *Works and Days,* Trans. Apostolos N. Athanassakis. (Baltimore: Johns Hopkins University Press, 1983), 110-95.

[56] Mumford, 77.

[57] Forbes, 4-5.

of bronze, were expensive, whereas iron was cheaper by reason of being "far more widely distributed over the earth's surface, so that far more people could obtain it for tools without having to organize elaborate transport and exchange." Iron tools became

> generally available to the farmer and enormously increased the productivity of agriculture. Before 1000 BC iron hoe-blades, plowshares, sickles and knives were in use.... From about 700 BC iron axes permitted the clearing of forests and allowed a great expansion of agriculture in Europe.

Dramatically increased surplus from agricultural productivity permitted communities to support large numbers of specialized craftsmen, whose products became generally available "instead of being the monopoly of the wealthy." Technological innovation, adaptation and specialization followed, for example the use of the pulley, the first evidence of which appears in an eighth century BCE Assyrian relief.[58]

Forbes points out that before engaging a discussion of the development of technology, it is necessary to understand the evolutionary process whereby material culture evolves, namely, "by the aid of *methods*, often dependent upon extraneous *means*, [humankind] employs *materials* for the achievement of *results*—not all of which are successful and survive as artifacts." Artifacts, however, arise through the "rough-and-tumble" of trial and error over time. Thus, at each stage of its development, metallurgy may have made its "first impression" as a curiosity from which evolved a need. Once the need was recognized, the accumulation of adaptions and dissemination quickly followed.[59]

By the time of written Greek, the word *technê* had acquired the sense of specialization in a productive field whose products were useful to the community and gave evidence of the skills or "craft" of the *technitēs*, the possessor of . By this time, the concept of *technê* also included an assumption of the systematic nature of the knowledge or skills involved, and their transmissibility by the *technitēs* through explanation and teaching.[60]

---

[58] S. Lilley, *Men, Machines and History* (London: Cobbett, 1948), 20-22.
[59] Forbes, 9-10.
[60] Roochnik, *Art and Wisdom*, 20-21.

David Roochnik points out significant shadings and shiftings of the term's meaning that emerged in Homer's *Iliad* and *Odyssey* as well as the later so-called "Homeric Hymns." Although he notes that the term generally applies to smithing, woodworking, or shipbuilding, it could equally be applied to the activities of the leather cutter or potter, as all the skills involved are useful, resulting in visible products. He characterizes what he terms the "Homeric conception" of *technê* by the following points:

1. A *technê* is directed toward a specific achievement, such as woodworking, smithing or weaving. It represents a bounded area of expertise which can be mastered and in which its master can be judged authoritative.

2. *Technê* is productive, i.e., skills associated with the possession of *technê* generally result in the creation of products: houses, ships, metalwork. Skills like piloting a ship, which do not result in material products, are also related to *technê*.

3. *Technê* has to be applied: the shipbuilder who merely possesses the knowledge but does not build ships is "not a complete shipbuilder." The *technitēs* is definitively identified with his field of production or craft.

4. The results of a *technê* are visible to the populace, and the community benefits from its products. Although the *technitēs* of Homeric Greece is not yet to be compared to the current-day professional who "hangs a shingle" advertising certification in a professional field, he is both in service to and uniquely recognizable within his community by virtue of his specialization.

5. *Technê* depends upon principles that can be rationally understood, and can therefore be explained, taught or otherwise transmitted.

Point 5 presumes an *intellectual* mastery of basic logic and cognitive skills such as rudimentary mathematics. This is a criterion that Roochnik believes represents a widening of the Greek concept of *technê* compared with evidence of its pre-Homeric meaning. He demonstrates that the pre-Homeric concept included the first four points. The addition of the fifth point accounts for a shift or widening of the meaning of the term to include the identification or association, in Homeric writings, of *abilities* (such as Proteus' shape-shifting) in addition to materially instrumental *skills* with *technê*. These skills have both an attitudinal and cognitive aspect.

A difference in the use of the term as it appears in the Homeric Hymns is that the meaning of the term begins to widen to include the abstract notion of planning or intellectual "craftiness." Roochnik points to the example of the *"technê"* of Proteus, whose skill is more of a doing than a making: the ability to shapeshift at will. And, "soon after Homer," Roochnik observes, *technê* "names skills, knowledges, both those like woodworking and smithing, which have tangible products, and those like lyre playing or the wiles of Prometheus, which do not."[61]

Roochnik mentions Homeric passages concerning "the prophet, doctor, singer or herald who...are said to be the *dēmiurgoi*, those who work for the *dēmos* rather than for themselves, and who are 'the men who all over the endless earth are invited,'" mentioned in Books 17 and 19 of the *Odyssey*. Although these are not yet identified with the *tektōn dourōn*, the possessor of materially productive skills, Roochnik points out the significance of the fact that they are the representatives of specialized accomplishments sufficiently recognizable to the *dēmos*, the community at large, that they should be invited (*klētoi*).[62] In other words, the doctor is recognizable as such and is welcomed to the community by dint of the known value of this type of skill.[63]

Solon (c. 600 BCE) uses the term *technê* only once (describing Hephaistos as *polutechneo*, or "master of crafts"), but as Roochnik demonstrates, it can be safely inferred that Solon introduces an ethical dimension in his reflections on the material foundations of human happiness and prosperity. In his "Prayer to the Muses," Solon reflects on the efficacy of crafts and the goodness of the gods' gifts of knowledge, which humans use to create wealth and happiness. He also urges his reader to reflect on the human propensity for self-deception in regard to the degree to which we actually control the consequences of our actions, including our exertions involving *technai*, regardless of how skillful or foresightful we may be.

---

[61] Roochnik, *Art and Wisdom*, 25-26.
[62] Homer, *Odyssey*, 17:386, qtd. in Roochnik, *Art and Wisdom*, 25.
[63] Roochnik, *Art and Wisdom*, 25.

"Indeed, it is *moira* [fate] that brings evil and good to mortals, and the gifts that the immortal gods give are inescapable. Surely there is risk in every activity, and no one knows, when a situation begins, how it will turn out."[64]

In this statement Roochnik infers the origin of the notion of the "value-neutrality" of *techne*. In his Prayer, Solon warns against seeking happiness through actions which are unjust (*adikōs*) or through what Roochnik translates as "lawless violence" (*hubris*). Such actions will draw the punishment of all-seeing Zeus. However, even those who possess skill and intend no ill may not always hope to see favorable results from their efforts, and good fortune is well known to come on occasion to those who lack good judgment and foresight. Though we can decide how we will dispose of the products of our *techne*, it is *moira*, not human will or instrumentality, that determines the outcome. Based on Solon's text, Roochnik adds a sixth, Solonic item to the list of five Homeric criteria for a *techne* given above:

6. "Possession of a *techne* leads only to a successful execution of a set of rational procedures; it does not lead to happiness or true human flourishing."

Roochnik believes that Solon's inference prefigures Aristotle's distinctions, in the *Nicomachean Ethics*, between *techne*, which is to be judged by the quality of its products, and ethical virtue, which involves "the more difficult task of evaluating the motives and character of the agent." By the time of the dialogues of Plato (428-348 BCE) and the writings of Aristotle (384-322 BCE), qualitative value resides in *techne*, but ethical value, the determinant of "goodness," resides exclusively in the disposition or intent of the possessor of technical knowledge.[65]

In Aeschylus' *Prometheus Bound* (written c. 460 BCE), Prometheus is chained to a cliff in punishment for aiding humankind. In a speech to the Chorus, Prometheus arrogates to himself credit for gifts to humankind for which the writer of the Homeric Hymns formerly praised Hephaistos and Athena:

---

[64] Solon, qtd. in Roochnik, *Art and Wisdom*, 28.
[65] Roochnik, *Art and Wisdom*, 30-31.

Men and women looking saw nothing, they listened and did not hear, but like the shapes in a dream dragging out their long lives bewildered they made a hodgepodge of everything, they knew nothing of making brick-knitted houses the sun warms, nor how to work in wood. They swarmed like bitty ants in dugouts in sunless caves.[66]

Prometheus claims credit for inventing and teaching to humans the arts of house-building and woodworking, animal husbandry, shipbuilding, medicine, and metallurgy, all reckoned since the time of Homer to be *technai*. Prometheus also mentions medicine and prophecy, *technai* of the *dēmiurgoi* whom Homer acknowledged to be rightly famous for their skills. Roochnik sees mathematics and language as a significant addition to this list of Promethean inventions.

Prometheus describes mathematics as "wisdom above all other." Although the other skills are *technai* in the traditional sense of being productive, either directly in the case of housebuilding or less directly in the case of medicine, mathematics is not. Philolaus (c. 470-c. 385 BCE), who is conjectured to have been an influence on Socrates, says that mathematics is fundamentally determinative (i.e., defined, defining, bounded, and replicable). If a *technê* is a bounded, determinate field of knowledge, then ultimate criteria must exist that can be called upon to judge the authority of the *technitēs* ("technician"). Philolaus adds that "there will not even be an object of apprehension at all if everything is indeterminate." And, "indeed all objects of apprehension have *arithmos* ["number"], for it is not possible for us to think of or apprehend anything without this."[67] In the *Republic*, Socrates observes that, "every *technê* and *epistēmē* [rational knowledge] is forced to participate [in] counting (*arithmos*) and calculation (*logismon*)."[68] Mathematics, then, is systematically intelligible, precise, and its explanatory function is definitive. Later, Aristotle will give mathematics as the prime example of knowledge which is "theoretical...i.e., one with no practical application at all."[69]

---

[66] Aeschylus, qtd. in Roochnik, *Art and Wisdom*, 33.

[67] Roochnik, *Art and Wisdom*, 37.

[68] Plato, *Republic*, 522c, *The Collected Dialogues of Plato*, eds. Edith Hamilton & Huntington Cairns, Bollingen Series 71 (Princeton: Princeton University Press, 1961).

[69] Roochnik, *Art and Wisdom*, 36.

Yet, by the 4th century BCE in Athens, it is understood by philosophers that the abstract system of mathematics must provide the criteria for describing and regulating material production.

Mathematics, then, as the paradigm of "intelligibility," would in this sense function as a paradigm for all *technai*.[70] Thus, based on how the term is used by Aeschylus and later authors, Roochnik presents the following list of criteria for *technê* according to the "Promethean Conception:"

1. "A *technê* (with the possible exception of writing) has a determinate subject matter or field."

2. "It has a useful result, either directly, as in the case of the shipbuilder, or indirectly, as in the case of writing and arithmetic, whose knowledge can be used to supplement and assist the process of production."

3. "It promotes human independence from gods, nature, and chance."

4. "It must be applied, either directly or indirectly, in order to become useful."

5. "It is easily recognized."

6. "The technai are teachable…. A *technê*, then, has a rational content, a *logos*, that can be communicated."

7. "Perhaps because of its exemplary intelligibility, arithmetic is paradigmatic."[71]

The "Promethean Conception" of *technê* differs from the pre-Homeric and Homeric conceptions given above in two points. First, the sense of recognition by the community at large is absent from this conception; instead, *technê* promotes independence from gods, nature, and chance. Second, the list now includes mathematics as paradigmatically definitive, and as distinct from typically productive *technai*.

By the time of the Platonic dialogues, Roochnik discerns a background question which is nevertheless central: "What is the

---

[70] Roochnik, *Art and Wisdom*, 39.
[71] Ibid, 41.

relationship between *technê* and *arête*?" *Arête* is "excellence" in a specifically moral sense, i.e., excellence in behavior in relation to the community. Is *arête* a teachable *technê*? Roochnik believes that the definitive answer to this question for Plato is "No." *Technai* are, by Plato's time, by definition teachable. Various dialogues present examples of *technai* that fail to produce recognizable *arête* in those who have learned them. "Socrates seems to be properly skeptical about the intrinsic benefit of any given *technê*; what finally matters is whether a man is law abiding and decent or not." In the dialogue *Laches*, "what matters is not whether a soldier masters the latest technical advances in warfare, but how brave he is." By the time of Plato, then, what is critical is that *technê* is to be seen as value-neutral. A *technê* can be applied for good or ill and cannot be said to influence moral excellence (*arête*). In the end, "there is a gap between a technical and a moral education."[72]

A significant shift occurred in the meaning of the Greek term *technê* between the time of Homer and Plato. Before Homer, *technê* signified skills resident in and practiced for the well-being of the community. Well after Homer's time, the *technitēs* was individually recognized within a community, and possessed an identity and was accorded respect in connection with his *technê*. Yet, the roots of depersonalized technology—of a separation of moral value, in both the personal and communal contexts, from the skills originally praised as the gifts of the god in the Hymn to Hephaistos—are clear in the Platonic dialogues.

## Mêtis

Hephaistos is paradigmatically *klutotechnês*, famed for *technê*. The concept of *technê* includes a systematization of knowledge and cognition. He is also paradigmatically *klutomêtin*, renowned for *mêtis*. *Mêtis* represents a different kind of intelligence or mental category, generally translatable as "cunning intelligence." Roochnik demonstrates how the concept of *technê*, though it describes a definitive relationship to matter and both divine and human instrumentality in regard to matter, nevertheless substantially

---

[72] Roochnik, *Art and Wisdom*, 90-92.

changed in meaning over the five centuries between its first literary appearance and its use by Plato. By contrast, in their etymological and mythological study of the "semantic field" of *mêtis*, Marcel Detienne and Jean-Pierre Vernant point out its "amazing and coherent stability throughout Greek history."[73] At the same time, it is indicative of a signal quality of *mêtis* itself that, where Roochnik can construct a systematic series of arguments tracing the development of the term *technê* in mostly linear fashion – in other words, there is, at least arguably, an identifiable *theōria* of *technê*, particularly by the time of Plato – this is not the case with Detienne and Vernant's findings with regard to *mêtis*. They assert the pervasive presence of *mêtis* "at the heart of the Greek mental world," but point out that its existence, though demonstrable, is not to be found in any definitive text; nor is it easily definable.

Detienne and Vernant identify *mêtis* as a way of knowing:

> a complex but very coherent body of mental attitudes and intellectual behavior which combine flair, wisdom, forethought, subtlety of mind, deception, resourcefulness, vigilance, opportunism, various skills, and experience acquired over the years. It is applied to situations which are transient, shifting, disconcerting and ambiguous, situations which do not lend themselves to precise measurement, exact calculation or rigorous logic.[74]

While Roochnik constructs a case demonstrating that, for the Greeks, by the time of the fourth century, the development of the concept of *technê* into a model for rational knowledge (*episteme*) leads to stability, measurability, objectivity, and singularity, Detienne and Vernant perceive in the concept of *mêtis* the characteristics of mutability, multiplicity, and obliqueness. *Technê* is stable. *Mêtis* moves; and its movement is an oscillation between the opposite conditions of stability and mutability. The god or mortal who dominates a situation through his or her *mêtis* does so by proving himself to be "more multiple, more mobile, more polyvalent than his adversary."[75]

---

[73] Marcel Detienne and Jean-Pierre Vernant, *Cunning Intelligence in Greek Culture and Society*, trans. Janet Lloyd (Atlantic Highlands: Humanities, 1978), 2.
[74] Detienne and Vernant, 3-4.
[75] Ibid, 5.

The mythological Metis is the first wife of Zeus. Zeus discovers that Metis is destined to become the mother of a son who will overthrow his father. So, when she becomes pregnant with Athena, Zeus swallows her. According to Apollodorus, Metis has the power of metamorphosis, and to elude the embrace of Zeus she changes form.[76] What we learn from this is two-fold: the nature of *mêtis* is paradigmatically shape-shifting, and Zeus, having swallowed Metis, is the paradigmatic possessor of *mêtis*.[77] This is of course how Athena, daughter of Zeus and Metis, comes to possess the shape-shifting ability that is key to her instrumentality in influencing human events, for example in the *Odyssey*, when she takes on the bodily form of Mentor in order to influence and guide Telemachus. We also learn that it is possible to confound the possessor of *mêtis*, but only through surprise and the trickery of ambush. Zeus must suddenly grasp and hold Metis through her transformations, and bind her inescapably; in this case by swallowing. It is through surprise and binding with an unbreakable bond that *mêtis* itself is overcome. Thus, the ability to capture and bind a wily moving target is key to the mastery of *mêtis*.

The nature of this binding is to be understood both literally, as in the case of a hunter's snare, and figuratively, in the sense of an opponent trapped by a feint into an unfavorable position, because of a belief that the adversary is holding a position opposite to where he truly stands. Detienne and Vernant illustrate this with the story of Antilochus in Book XXIII of the *Iliad*. Antilochus is at a disadvantage in a chariot race. His horses are slower than his opponents'. His father, Nestor, the wise sage possessed of *mêtis*, advises him to take advantage of a narrowing in the track, and to drive his horses suddenly in front of his rival, Menelaus, in order to surprise him by the very rashness of the action, which puts both drivers at risk of crashing. Antilochus accomplishes this and gains the advantage necessary to win the race.

This story shows that contests may be won by *mêtis* where sheer strength is lacking. Second, vigilant watchfulness is necessary to be prepared to take advantage of the opportunity for surprise or trickery.[78]

---

[76] Ibid, 20.
[77] Detienne and Vernant, 21.

Thus, the possessor of effective *mêtis* must hone his perceptions. A third feature of *mêtis* is its diversity and multiplicity.[79] Antilochus has further tricks up his sleeve by which he increases the odds that he will succeed by employing Nestor's strategy. Antilochus, who though young in fact possesses the *technê* of handling horses and chariot to a high degree of mastery, feigns impetuosity and immaturity. He fools Menelaus into believing he is not in control of his horses, diverting Menelaus' attention into urging him to take care, while Antilochus quite deliberately and with prudent forethought and consummate skill, quite the opposite to Menelaus' expectation, drives his horses into the breach.[80] This demonstrates a fourth feature of *mêtis*: that its very essence is illusion and deceit.[81] The story also shows that to be effective, *mêtis* assumes the possession and mastery of a *technê*.

In its essence, Detienne and Vernant identify two signal metaphoric characteristics of *mêtis* and of those possessing it. The polyvalency of *mêtis*, which inspires in its possessor both suppleness and obliqueness, can be represented metaphorically by *a reversed, circular, or skewed* gait.[82] This circularity is also expressed in the metaphorical *snare* set by the possessor of *mêtis*. The snare encircles the victim unaware and is suddenly drawn tight, binding the victim in an inescapable circular bond. Together, these metaphors describe the archetypal image of Hephaistos.

Perhaps the most distinctive epithet of Hephaistos is *periklutos amphigueeis*, or "the renowned cripple" (*amphigueeis* can also mean "ambidextrous"). The gait of Hephaistos is an emblem of his *mêtis*. In both Homer's and Hesiod's version of the god's origin, Hephaistos is born crippled, perhaps clubfooted, a defect which causes Hera to cast him from Olympus in horror and rage. In Homer's *Iliad*, Hephaistos falls a second time when Zeus ragefully casts him from Olympus when he takes his mother's part in challenging Zeus's absolute authority. He lands broken upon Lemnos, where the Sintian people rescue the god and nurse him to health but not to wholeness. I will later show how Hephaistos' lameness is mythopoeticized into an

---

[78] Ibid, 13-14.

[79] Ibid, 18.

[80] Ibid, 22.

[81] Ibid, 21.

[82] Ibid, 6.

imputation of sexual deficit, specifically symbolic phallic impairment or castration. J. P. Vernant and Pierre Vidal-Naquet, however, suggest an alternative interpretation of Hephaistos' crippled condition. The "divergence in the direction of his feet, *a gait oriented in two directions at once,* forward and backward," is "connected with his powers as a magician."[83] Hephaistos' special circular lameness is echoed in the magical wheeling tripods he fashions, described by Homer in Book 18 of the *Iliad* as "animated automata [that] could move forward and backward with equal ease." This circularity is emblematic of that which is fabulously and powerfully magical. Vernant and Vidal-Naquet observe that,

> The movement of Hephaestus, the lame god, "rolling" around his bellows in the workshop, was circular.... So was that of the primordial men, those beings described by Aristophanes in the Symposium who were "complete" in comparison to the men of the present, cleft as they are in two (down an axis separating front from back).[84]

Vernant and Vidal-Naquet also remark that in describing the men suited to form the elite ruling class of the ideal Republic Plato compared those whose thinking is "deformed and lame," which is to say divergent from the linear rational mode of Platonic philosophy, to those whose patriarchal line of descent is skewed or deformed. In short, he compares those whose thinking is divergent—i.e., characterized by the qualities of *mêtis*—to bastards and illegitimates. What makes individuals "well-born" and qualified to rule is the straightness and linearity of their thinking as much as the excellence and legitimacy of their lineage.[85] In this light, it is interesting to note that Hephaistos, the crooked-gaited god of technology, had, at least in Hesiod's account of his parentage, a mother but no *father.* It is also here that the definitive philosophic fracture between *technê* and *mêtis* can first be seen.

Circularity or bidirectionality as opposed to straightness of gait is one of the clues Detienne and Vernant use in capturing the elusive qualities of *mêtis* itself, and in discerning who among the gods

---

[83] Jean-Pierre Vernant and Pierre Vidal-Naquet, *Myth and Tragedy in Ancient Greece*, trans. Janet Lloyd (New York: Zone, 1990), 98, italics mine.
[84] Vernant and Vidal-Naquet, *Myth,* 210.
[85] Ibid, 211.

possesses *mêtis* and who does not. Hermes, who has *mêtis*, steals the cattle of Apollo, who does not. Hermes accomplishes his trick of making Apollo's cattle disappear by driving them before him while wiping out and then reversing their tracks and himself walking backwards. The same source ("The Homeric Hymn to Hermes") also describes Hermes compelling the cattle to walk backwards, their heads turned toward the thief leading them, while Hermes himself "adopts a 'twisted gait' (*epistrophádēn*)."[86]

As Hephaistos' gait is symbolic of the circularity of *mêtis*, so is one of the most famous of Hephaistos' creations symbolic of the snaring or binding character of *mêtis*: the golden net already mentioned that was created by Hephaistos as a snare to catch Ares and Aphrodite in the act of adultery. As in the example of Antilochus, Hephaistos combines forethought, consummate skill, and cunning trickery to take his revenge. He is informed of the adultery of Ares and Aphrodite by all-seeing Helios, who has spied them in Hephaistos' bed. He rages, but his rage never displaces his prudence nor his consummate command over his *technê*. He makes a plan to trap the lovers and executes it. First, he limps into his workshop and creates a net of unbreakable chains, "bonds that no one can loosen."[87] Then he lays them in a circular pattern around the bed and suspends the ends from the ceiling. The net of chains is "gossamer-fine as spider webs no man could see, / not even a blissful god...."[88] Next, he lightly tells Aphrodite he is taking a trip to Lemnos, both lying and adopting a false face in a similar manner to Antilochus' ploy of pretending inexperience. Ares and Aphrodite literally fall into the snare, the timing perfect for Hephaistos to limp in to the bedchamber, having called the other gods to witness. Like Antilochus winning the race against faster horses, so slow-moving Hephaistos has used his wits to catch the fastest-running of the gods.

Significantly, Hephaistos also outwits Aphrodite, who does possess *mêtis*, demonstrating a point made above, that the possessor of *mêtis* may outwit even another who possesses it, by being "more

---

[86] Detienne and Vernant, 301.

[87] Ibid, 284.

[88] Homer, *The Odyssey*, trans. Robert Fagles (New York: Penguin, 1996), 8:318-9.

multiple, more mobile, more polyvalent than his adversary" and imposing unbreakable, constraining bonds upon the adversary.[89]

Roochnik asserts that, to the degree to which Detienne and Vernant connect *mêtis* with *technê* in their analysis, "What they fail to take into account is how the meaning of '*technê*' evolves and eventually...refers to a kind of knowledge much more stable and rigid than the flexible shrewdness of *mêtis*."[90] Roochnik presents evidence clearly identifying a pivot point on which the meaning of *technê* evolves away from an emphasis on the outcome of a physical product based on the systematic application of craft, toward an emphasis on systematic and verifiable *thinking* as a necessary precursor to the products of craft. This then also makes possible the evaluation of these products against objective standards of measurement, hence objective excellence.

I suggest that Detienne and Vernant are not to be faulted with connecting *mêtis* with *technê* in the case of Hephaistos. To the contrary, as will be seen in the discussion of the image of the blacksmith-creator gods (in Chapter 4), the Hephaistos archetype is a container of seeming opposites, in this case both mastery of craft as definable by standards of quality and mastery of oblique strategies that take a nonlinear, circular route. The image of the golden net is at once emblematic of Hephaistos' *technê* and his *mêtis*. It will be seen that the art and craft of Hephaistos do not rely on *technê* alone, and that the fragmentation of the Hephaistean mythos can, at least in part, be traced to what Roochnik identifies as the definitive parting of ways between craft and 'craftiness' in the development of the concept of *technê*.

## *Poiēsis*

The Greek word *poieô* signifies making, creating, bringing material products into existence. Hephaistos the blacksmith is the god of *poiēsis*, "making." The "Homeric Hymn to Hephaistos" credits the god with the making of habitations that make the quality of human life good, separating mortals from the beasts who dwell in caves.

---

[89] Detienne and Vernant, 5, 284.
[90] Roochnik, *Art and Wisdom*, 24 n.14.

Hephaistos not only quells the threat of a fearful fight between Zeus and Hera at the beginning of the *Iliad*, but provides the place where each god retires peaceably afterward: "At last...each immortal went to rest in his own house, the splendid high halls Hephaistos built [*poiêsen*] for each with all his craft and cunning, the famous crippled Smith."[91]

Images of artifacts fashioned by the god of fire and forge teem in Homer. In the *Iliad*, they are mostly implements of war. Apollo appears among the Trojan warriors to lead a charge,

> shoulders wrapped in a cloud, gripping the storm-shield,
> the tempest terror, dazzling, tassels flaring along its front—
> the bronzesmith god of fire gave it to Zeus to bear
> and strike fear in men...."[92]

Not gods alone but mortals bear arms of Hephaistos' making. Diomedes, among the fiercest of the Achaeans, wears armor forged by Hephaistos.[93] Most famous of all is the armor—shield, breastplate, greaves and plumed helmet—created by Hephaistos for Achilles at the request of his mother Thetis. The glare alone of the armor, "burnished bright, finer than any mortal has ever borne across his back," is enough to cause Achilles' Myrmidons to tremble in their ranks, shrinking from the very sight of it.[94]

Another significant class of objects introduced in the *Iliad* consists of emblems of royal legitimacy and power. When, in Book 2 of the *Iliad*, King Agamemnon angrily denounces Zeus as the cause of the Achaeans' failure over nine long years to breach the Trojan walls, he brandishes his royal scepter as emblem of his temporal power. This scepter is of Hephaistos' making. Homer traces its passage via gift from its maker to Zeus, from Zeus to Hermes, Hermes to Pelops, Pelops to the grandfather of Agamemnon, from whom Agamemnon has inherited it as emblem of the kingly power of the House of Atreus.[95] Thus, King Agamemnon raises the very scepter that once belonged to Zeus to denounce the meddling of the king of heaven in men's affairs.

---

[91] Homer, *Iliad*, trans. Fagles, 1:728-31.
[92] Homer, *Iliad*, trans. Fagles, 15:362-66.
[93] Ibid, 8:220-21.
[94] Ibid, 19.12-18.
[95] Ibid, 2:118-65.

Already mentioned above as a key example of Hephaistos' making is the golden net he fashions to snare Ares and Aphrodite, which Homer calls "a masterwork of guile."[96] A net so fine as to be invisible even to the eyes of the gods is a magical one. The story of Hera's throne reveals a similarly tricky, secretive, and magical mytheme embedded in the catalogue of Hephaistean creations. Timothy Gantz traces the earliest mentions of this story about Hephaistos' trick of revenge against his mother to the sixth century. Later writers supply further details.[97] When Hera sits in the magnificent throne, it flies magically upwards. At the same time, manacles bind Hera fast to it. Some versions describe her hanging helplessly suspended, upside down. None of the Olympians (who, it should be noted, generally lack the direct instrumental powers over matter commanded by Hephaistos) is able to rescue her.

Another instance of Hephaistos' magical ability to animate objects of his creation is found in the *Iliad*, in the description of his workshop. The crippled, hobbling smith orders his bellows, which clearly possess their own degree of *technê*:

"Work—to work!"
And the bellows, all twenty, blew on the crucibles,
breathing with all degrees of shooting, fiery heat
as the god hurried on—a blast for the heavy work,
a quick breath for the light, all precisely gauged
to the god of fire's wish and the pace of the work in hand.[98]

The hobbling smith is assisted in the forge by capable handmaids of his own making:

all cast in gold but a match for the living, breathing girls.
Intelligence fills their hearts, voice and strength their frames,
from the deathless gods they've learned their works of hand.[99]

The handmaids also recall Hephaistos' creation of Pandora. This story is found in both Hesiod's *Works and Days* and his *Theogony*.[100] Zeus, angered that Prometheus has placed the divine gift and

---

[96] Homer, *Odyssey*, trans. Fagles, 8:318-19.
[97] Gantz, 75.
[98] Homer, *Iliad*, trans. Fagles, 18:548-53.
[99] Obod, 18:488-91.
[100] Gantz, 78.

prerogative of fire into the hands of men, orders Hephaistos to "mold together earth in the form of a maiden." Hephaistos gives her "voice and strength," and "liken[s] her in face to the goddesses."[101] The other gods then give her various "gifts," which is why her name reflects the participation of *pan-*, "all." Gantz notes that some versions of the story mention hammers being involved in Pandora's making, "which might help to explain a blacksmith god's involvement."[102]

In tracing the origins of the notion of craft and craftsman in connection with the development of smithing and its meanings in folklore, Lotte Motz explores the significance of ceramic technology in Mesopotamia which in Neolithic times had already reached a high degree of sophistication and competence. Ceramic technology requires the ability to control high-temperature firing in ovens and knowledge of the chemical properties and reactions of pigments in the process. The technology developed by the potters was later adapted to the processes of bronze and coppersmithing.[103] The reason this may be significant, Motz argues, is that this earlier technology may offer insights into the attribution of mysterious powers to the smith in his transmutation of crude matter into powerful implements.

Motz points to the numinous significance of clay figurines or statues and myths of humans created from clay. Sumerian Nimah "created man from earth with the help of 'good and princely fashioners.'"[104] Arruru, the goddess of creation, made Enkidu from clay; Yahweh formed Adam; the Egyptian god Knum formed men on his potter's wheel, and of course Hephaistos made Pandora, the first woman, from clay or earth. In some languages, Motz observes, clay and earth are designated by the same word, as in Germanic "earth" and "earthenware," Latin *terra* and Hebrew *admah*.[105] Clay and earth are sacred substances.

Clay and stone were early incorporated into ritual activities. The carved clay figurines described by Marija Gimbutas as manifestations of the Great Mother appear as early as 35,000 BCE, carved from

---

[101] Ibid, 155-56.
[102] Gantz, 78.
[103] Motz, 142.
[104] Ibid, 147.
[105] Ibid, 148.

nonorganic substances or formed of fire-hardened clay. Clay animal figures discovered in the Montespan cave in the Haute Garonne in the French Pyrenees have wounds modeled into them, much as cave drawings elsewhere show darts and arrows in association with animal figures, strongly suggesting the use of sympathetic magic to ensure the success of the hunt. Clay effigies of women and animals as well as life-sized statuary were common in seventh millennium BCE Jericho, and clay figures of women were also found together with painted clay bowls and jars in Hacilar in Anatolia. In north-western Germany in the Neolithic period, pottery vessels formed the largest part of grave-gifts in tombs. Before the time of metalsmithing, images of stone axe blades were incised in megalithic sepulchers in Western Europe and votive axes of stone and amber are present in northern burials. The double axe fashioned of metals later becomes a prominent religious symbol. In these and numerous other examples such as standing stones, corbelled vaults and brick mortuaries, Motz finds that "The works of the pre-metal craftsman, the potter, builder, and stone smith…were vital to the human effort of finding a pathway to the gods."[106]

Though later forged in the smithy, the thunderbolts of the early sky-gods were originally imagined as and retain in their etymology the memory of stone. Motz cites the English term "thunder-stone," applied to flint celts (hafted blades), the German terms *Donnerstein* and *Donnerhammer*, French *pierre à tonnère* and Greek *keraunia lithos* as evidence. Likewise, Thor's hammer, forged by dwarves, was originally stone as shown by its name, Old Icelandic *hamr*, meaning "rock" or "precipice." Magical stone cudgels named "Crusher" and "Driver" were made by Kothar-wa-Hasis for Baal. These and other examples are given by Motz to show that "throughout the confines of Europe [and elsewhere]…stone celts and axes are regarded as the thunderbolt."[107] Numinous power residing in stones and later reflected in sacred and funeral architecture is perhaps most simply shown in the planting of a boulder in the earth by Jacob, who named it Beth-El, the house of God.[108] This numinous power was later

---

[106] Motz, 143-45.

[107] A. B. Cook, *Zeus: a Study in Ancient Religion*, qtd. in Motz, 147.

[108] Motz, 145.

translated to metal and metalworking, for example in beliefs such as that in Scotland, where the possession of a piece of iron, or a knife or nail carried in one's pocket might protect one from fairy mischief, or in India, where the belief that a mourner attending to the dead might ward off evil spirits by carrying a piece of iron.[109]

With the grounding of Motz's work in uncovering the most ancient derivations of ideas concerning matter and its numinous significances, it becomes easier to understand Heidegger's definition of *poiēsis* as a "bringing-forth" (in German, *hervorbringen*) of the inherent message of matter in relationship with the human maker. Heidegger points out the distinction between the natural bringing-forth of nature, *physis* (the root of our "physics"), which, like a blossom, is "the arising of something from out of itself" and the bringing-forth of an object like a silver chalice through an *other*, i.e., the craftsman or artist. Yet, by calling the bringing-forth of nature "*poiēsis* in the highest sense," Heidegger does not thereby diminish the human maker. The point, instead, is that *poiēsis*, whether a work of nature or of handicraft, reveals something, "brings hither out of concealment forth into unconcealment." That unconcealment of something already present in its essence the Greeks called *alēthea* which means "revealing." "The Romans translate this with *veritas*. We say 'truth' and usually understand it as the correctness of an idea."[110]

This bringing-forth/unconcealment/revealing/bringing-into-being is accomplished through instrumentality, i.e., *technê*, so that its presence shines forth.[111] "Technology is therefore no mere means. Technology is a way of revealing." The essence of technology lies in "the realm of revealing, i.e., of truth."[112] Heidegger looks further into the Greek word to clarify the idea of "truth" as intimately connected with making. The Greek word *theōria* (from which we have "theory") from the roots *thea* and *oraō*, means "to look attentively on the outward appearance wherein what presences [that which has come into being] becomes visible and, through such sight—seeing—

---

[109] Ibid, 143.

[110] Martin Heidegger, *The Question Concerning Technology and Other Essays*, trans. William Lovitt (New York: Harper, 1977), 10-12.

[111] Heidegger, 164.

[112] Ibid, 12.

to linger with it." Not merely a mode of thought, *theōria* signifies the attentiveness of humans to the radiance that lies within the outward appearances of things, in which the presence of the gods shines forth. For the Greeks, this attentiveness was "the consummate form of human existence." Heidegger stresses that the Greeks

> were also able to hear something else in the word *theōria*. When differently stressed, the two root words *thea* and *oraō* can read *theá* and *ōra*. Theá is goddess. It is as a goddess that Alēthia, the unconcealment from out of which and in which that which presences, presences, appears to the early thinker Parmenides.... The Greek word *ōra* signifies the respect we have, the honor and esteem we bestow.... [Thus,] *theōria* is the reverent paying heed to the unconcealment of what presences. Theory in the old, and that means the early but by no means the obsolete, sense is beholding that watches over truth. Our old high German word *wara* (whence *wahr*, *wahren*, and *Wahrheit* ["truth"]) goes back to the same stem as the Greek *horaō*, *ōra*, *wora*.[113]

Thus, *technê*, technology, "is the name not only for the activities and skills of the craftsman, but also for the arts of the mind and the fine arts. *Technê* belongs to bringing-forth, to *poiēsis*; it is something poetic." It is sacred.

Heidegger acknowledges that there is a disturbing problem with applying these ideas to the essence of modern, mechanistic technology. Technology is, as it always was, a revealing. However, "The revealing that lies in modern technology is a challenging, which puts to nature the unreasonable demand that it supply energy that can be extracted and stored as such.... The earth now reveals itself as a coal mining district, the soil as a mineral deposit."[114] The worldview that feels the presence of *Alēthia* as goddess and that reverently observes the numinosity of the body of nature is lost.

In her examination of the smith, Motz goes back to very ancient traditions that render an understanding of *poiēsis*, "making," as partaking in the numinous. Matter is implicated in an interplay with divine forces; both making and matter are rendered magical. Technology and art are makings that reveal the essence of what lies under material reality and shines forth: whether the making is a silver chalice or a poem, a "truth" shines forth through the agency of the

---

[113] Heidegger, 163-65.
[114] Ibid, 13-14.

maker. And this agency is significant, for without it, the potential of things remains hidden, and cannot be known. In Chapter 5, I will give examples of "Hephaistean" technologies that honor nature through making that is both reverential and productive of "good things for life."

## Ekphrasis

This dissertation characterizes the Greek god Hephaistos as the embodiment of the archetypal image of making. I have shown the connection between *poiēsis* (making) and *technê* (that which pertains to the skill of the maker), which reveals the truth (*alēthea*) of nature and the divine. Handcraft manufacture, art, and poetry all bring forth this hidden truth. Thus the *poiēsis* of the blacksmith and the poet reveals truths through application of *technê*.

But how is this so? In order to reveal the truth hidden in matter, waiting to shine forth, the craftsman, the artist, the poet, not only acknowledges himself to be in *relationship* with what is hidden but also to participate in the *process* of creation whereby it is revealed. There is a riddle in creation whereby the revealed creation itself is subject to the reciprocal gaze of the observer, and that something well and rightly made continues to speak its truth, to be continually uncovered. Thus the deeper meaning of *poiēsis* is many-layered and dynamic—for the Greeks as indeed for us. This idea will be conveyed best through example.

In describing the artistic activity of Hephaistos in making the Shield of Achilles Homer is not only presenting the object made by the god to the imaginal eyes of his readers/listeners, he is calling our attention to the divine nature of its making. At the same time, he is calling attention to his own divinely inspired making in doing so.

A significant part of the poet's *technê* is the art of *ekphrasis*. This is a term that has been taken into English literary terminology and has been the subject of ongoing discussion. Its current meaning is "description," but its import is considerably broader in the context of attention, intention, and making. The paradigmatic example of *ekphrasis* in literature is Homer's lengthy and detailed appreciation of the making of the Shield and the marvelous scenes depicted on it.

Andrew Sprague Becker has devoted an entire scholarly work to Homer's *ekphrasis* on the Shield. In his book, Becker directs his reader's attention to the opening words of the *Iliad, mênin aeide theâ*: "Sing the wrath, goddess." In the very first word, Homer tells his listener and reader first that the subject of the *Iliad* is *mênin*, wrath (the wrath of Achilles, son of Peleus). With the second word, *aeide*, he calls attention to the medium in which the story will be conveyed, alerting us "that we will not experience 'wrath' but an artistic rendering of wrath in song."[115] Note that Homer does not use the noun *aoidê*, "song," but the "verb of creation," *aeide*, "sing." This, says Becker, "tells the audience that the work of verbal art is not to be imagined as a completed, preexisting product; it is a process, and it will be created as one listens."[116] More, the verb is presented in the imperative mood, calling attention to the presence of the Bard himself, "who is both source (for the audience) and audience (for the Muse's performance)." The *Odyssey* begins with a similar structure, *andra moi ennepe, Mousa*: "Tell me [of] the man, Muse." Again, the first word names the subject of the work, the man (Odysseus). In this case, the second word *moi*, "me," announces the presence of Homer the poet, who addresses the Muse on behalf of the audience.

This is a gentle operation of rhetorical *mêtis* whereby the Bard announces himself, not as the creator of the *Iliad* and *Odyssey*, but as its first listener, a mediator who is intimately linked both to the goddess and to the audience. He asserts both his own authority and at the same time presents himself as a reliable guide to a story for whose origin the poet himself is not responsible, but as one whose faithful representation of the story the audience can trust. What the audience will receive are the poet's transmission of as well as his reactions to the story he receives from the Muse. His reactions will often guide the trusting audience in how to respond and what emotions to feel. For example, in the second half of the *Iliad*'s first line, which enlarges on the "wrath" that is the subject of the story, Homer tells his listeners that this wrath is "destructive," that it "brought upon the Achaeans countless pains."[117] Notice that it is not the Goddess who tells us this. It is the poet.

---

[115] Andrew Sprague Becker, *The Shield of Achilles and the Poetics of Ekphrasis* (London: Rowman and Littlefield, 1995), 45.

[116] Ibid, 45, italics mine.

In classical *progumnasmata*, or handbooks on rhetoric (the earliest extant is from the first century CE), the term *ekphrasis* denoted the specific poetic skill of "bringing that which is being made manifest vividly before the sight." The rhetoricians specified that in order to accomplish this vividness, the language of description must "imitate completely the thing being described," and should "fit the form of the narrative."[118] Further, the description should eschew explicit interpretation. "The narration of the subject matter is bare," diminishing attention to the medium (the poem) and the describer (the poet).[119] At the same time, the earliest known of the handbook authors, Aelius Theon, also urges the student to follow the example of Homer in expressing reactions to what is described, beginning "from the fine and useful and the pleasurable, as Homer did in the arms of Achilles, saying that (they were) fine and strong and astonishing to see for his allies, but fearsome for the enemies." Registering the astonishment of Achilles' allies and the fear of his enemies at the sight of the fine arms and shield made by Hephaistos reminds the audience "of its own mediated access to the described phenomena," and of the position of the poet between the audience and that which is described.[120] Further, the poet is guiding the reader's emotional response to what is described.[121] But, the reader of *ekphrasis* is not meant to be imaginatively restricted to the author's view. The handbooks also state that the effect of the poet's style should "*almost* produce sight through hearing," "*all but* making us spectators." The enchantment is never meant to be quite complete: the *conscious* participation—and imaginative interaction—of the audience is implicit.[122]

Becker believes that Homer and any effective author of *ekphrasis* not only describes phenomena but instructs the reader in how to read the literary work. Becker summons the work of Paul Ricoeur (from his essay "Appropriation") to explain two categories of response to

---

[117] Ibid, 46-47.
[118] Ibid, 25-26.
[119] Ibid, 27.
[120] Ibid, 28-29.
[121] Ibid, 34 n. 64.
[122] Ibid, 28.

literature, which Ricoeur names "divestiture" and "acquiescence," as follows:[123]

| *Divestiture* | *Appropriation* |
|---|---|
| —Acceptance of illusion | —Attention to the working of illusion |
| —Enchantment | —Self-consciousness |
| —Literature as escape or diversion | —Literature as "equipment for living" |
| —Context of creation | —Context of reception |
| —Literature as experience of foreignness, difference | —Literature as experience of ourselves, recognition |

The reader of an effective *ekphrasis*, then, is invited, through the skill of the poet, into both a divestiture of and an appropriation of the text.

Echoing the first line of the *Iliad*, the first line of the *ekphrasis* on the Shield "contains a verb of fashioning:" "On [the shield] he made [*poisê*] two cities...." Like the singing of the Muse, the verb *poisê* is a word *denoting process rather than product*. It calls attention to Hephaistos as the maker of the Shield, just as the Muse is identified as the source of the poem. The presence of the Bard is reasserted in the second line, when he calls the two cities depicted on the Shield "beautiful," directing the audience's response while reminding us that the Shield is not visible to our physical eyes, but is visible to our imaginal eyes though the efforts of the Bard.[124] Becker asserts that the parallel structure of the first lines of the poem and the first lines of the *ekphrasis* "creates, in our experience of the poem, a consonance between life and depicted life." It

> diminishes the audience's concern for the mediating quality of the visual arts; the techniques of mimesis are noted, appreciated and

---

[123] Ibid, 39.
[124] Becker, 48.

accepted, while the description elaborates the referent. The description will supplement the visual image, going beyond what could be seen, but this is neither a statement of mimetic primacy nor a questioning of the representational capabilities of the visual arts.[125]

Rather, it reflects the poetics of Homer's work; and it reflects the poetics of Hephaistos' making.

It is not merely through description but through Homer's poetics that the audience is led to appreciate the divine and magical lifelikeness of Hephaistos' work. The beginning of the *ekphrasis* on the Shield is preceded by the appearance of Hephaistos' attendants, the "handmaids" who "rushed to help their master." In the next line of Homer's poem, the fact is revealed that they are "golden, resembling living girls."[126] The manner in which they are introduced gives a doubled motion to the image of representations endowed with mobility. They rush—and, they are golden! They are not living, they are made, but they doubly resemble living girls through their form *and* their motion.[127] On this basis, Becker asserts that the gradual additions of vivification and even sound in association with the images depicted on the Shield embody in poetic structure and language the quality of the mimesis achieved by the god—his replicas achieve "the ultimate impulse of representational art"—to transform the "replica into a living original."[128] In this way, the poem itself becomes a *mise en abîme*. While reminding the audience that although,

> art is not life, it [the artful *ekphrasis*] serves to increase our admiration for the visual art...., Celebration of the process, of what art can co, rather than a need for illusion or struggle for mimetic primacy, characterizes the mode of mimesis in the *Iliad* and specifically the Shield of Achilles.[129]

Becker provides many detailed examples of the recurrence of echoing, patterned structures that serve to both distance and involve the audience in regard to the "reality" being described as well as the poet's art in presenting it. The term and rhetorical concept of

---

[125] Ibid, 50.

[126] Homer, *Iliad*, trans. Fagles, 18:418-19.

[127] Becker, 80.

[128] Ibid, 82.

[129] Ibid, 85.

*ekphrasis* (engaging description of a made thing) provides an example intended to illustrate the concept of *poiēsis* (making) as it will be developed in this dissertation. Although Homer's *technê* can be evaluated in the abstracted sense in which it came to be understood by the time of Plato (and, indeed Becker's work is such an evaluation) the *mise en abîme* that is constructed around the reader's increasing enmeshment in the experience of the poem is more comparable to the invisible net with which Hephaistos snares Ares and Aphrodite. Once dropped, the net becomes visible and the quality of its craft can be appreciated by all participants in the presentation. Yet, the participants still own their participation: Apollo and Hermes can laugh; Ares and Aphrodite, when released, may take flight. The image of the maker, as understood through the reflexive lens of the Hephaistean archetype, includes the maker's implicit bond with the source of the object (this includes the creative urge or inspiration as well as the inherent nature of the maker's medium), with the process of making, and with the viewer/audience/user of the made object. Its truths continue to be uncovered.

## Mythos and Logos

To understand the uniquely Hephaistean archetypal connections of *mythopoesis* as creative activity, and their implications for re-imagining art and technology in the contemporary mind, it will be useful here to examine the origins and alterations in the meaning of the Greek term *mythos*, myth. Like the term *technê*, discussed above, the term *mythos* underwent significant change in meaning between the time of Homer and that of Plato. It is important to understand these changes both as a precursor to examining the concept and creative possibilities of *mythopoesis*, as well as to further establishing evidence for the historical frame as well as the nature of what I have chosen to term the "fragmentation" of the Hephaistean mythos.

It has long been noted by scholars that a "transformation" in Greek speech and thought occurred that "led from the *mythos* of Homer and Hesiod to the *logos* of Heraclitus and Plato."[130] However, as Bruce Lincoln's textual and etymological research painstakingly

---

[130] Bruce Lincoln, *Theorizing Myth: Narrative, Ideology, and Scholarship* (Chicago: University of Chicago Press, 1999), 3.

demonstrates, it is not a transformation in the sense of a transition from valuing one mode of thought or expression to another. Instead, it can be seen as a gradual and complex transposition in meanings between a set of two existing terms and concepts.

In the proem to his *Theogony* 27-28, Hesiod (eighth century BCE) claims to be directly addressed by the Muses as follows:

> We know how to recount (*legein*) many falsehoods (*pseudea*) like real things, and
> We know how to proclaim (*gerusasthai*) truths (*alēthea*) when we wish.[131]

As Lincoln points out, two modes of speech are employed in these lines. The Muses "recount falsehoods" and "proclaim truths." The formula "falsehoods like real things" also appears in Homer's account of Odysseus' conduct when, though physically disguised by the goddess Athena, he deliberately hides his identity from his wife Penelope while evoking her sympathy through intentful words and actions.[132]

The term *logos* appears five times in Hesiodic texts, in three instances modified by the adjective *haimulios* ("seductive"), and in three instances associated with the term *pseudea* ("falsehoods"). Hesiod describes Pandora as the prototypical woman, "into whose breast Hermes placed 'falsehoods, seductive *logoi*, and a wily character.'"[133] Hesiod warns against succumbing to woman's guile:

> Do not let a woman with swaying hips deceive your mind.
> Seductive [*haimula*] and cajoling, she's seeking your granary:
> He who puts his trust in a woman, puts his trust in thieves.[134]

Homer tells of Calypso beguiling Odysseus with "soft and seductive *logoi*" to make him forgetful of Ithaca.[135]

The sinuous and seductive power of persuasion that allows a weak woman to overcome the greater physical strength of a male is not however the province of the female alone. Like Odysseus, certain men and gods also possess the skill of influencing through the use of

---

[131] Lincoln, 3 n.1.
[132] Ibid, 4; 4 n.3.
[133] Ibid, 6.
[134] Hesiod, *Works and Days,* 373-75; Lincoln, 6.
[135] Homer, *Odyssey*: 1:55-57; Lincoln, 9.

seductive *logoi*. Hermes is "seductive in his cunning" and tricks
Apollo with his "crafts and seductive *logoi*."[136] The term used for
Hermes' seductive cunning is *haimulo mêtis*. Hermes is of course
famed for his *mêtis*. However, it is Zeus who, as Lincoln points out,
having swallowed Metis and thereby internalizing the voice of
feminine cunning, secures his sovereignty through adding the quality
of feminine strategy to masculine force.

So, in these early texts, the speech of *logos*, whether employed by
male or female, is both seductive and duplicitous. When applied to
men, the term also connotes the opposite of the heroic, masculine
attitudes of battle lust and courage. Rebuking Menestheus, leader of
the Cephallenians, for failing to respond to a call to battle,[137]
Agamemnon accuses him of both cunning and cowardice in battle:
"You who are surpassing in evil guiles [*kakoisi logoisi kekasmeme*],
wily of spirit, / Why do you stand by, cowering in fear? Why do you
wait for others?"[138] When Eurypylus is wounded in battle,[139]
Patroclus provides two kinds of healing, administering herbal balms
and entertaining him with words (*eterpe logois*); in other words,
telling him stories to soothe his pain. But when he sees that action has
suddenly and urgently resumed, Patroclus breaks the mood of
tranquility by slapping his thigh in distress, "And the voice that
entertained breaks into a harsher, but also a more realistic, speech...:
'Eurypylus, I can no longer stay here, notwithstanding your need. A
great struggle has arisen.'"[140] The soothing, even healing, qualities of
*logos* must give way to the preeminent masculine business of battle.
In these examples, Lincoln demonstrates that in the texts of Hesiod
and Homer the term *legein* (the speech of *logos*) is associated with
deceptive, "crooked" speech—the speech of *mêtis*—and the poetic
falsehood inspired by the Muses.

By contrast, the speech of *mythos*, though it may be crude,
forceful, and lacking in tact or charm, is true speech. The verb used in
Hesiod's *Theogony* to denote the proclaiming of truth by the Muses,
*gērusasthai*, is also found in his *Works and Days*. There it denotes the

---

[136] "Homeric Hymn to Hermes," 317-318; Lincoln, 9.
[137] *Iliad*, 4:339-40.
[138] Lincoln, 9.
[139] *Iliad*, 15:390-400.
[140] Lincoln, 10; 10 n.34.

speech of the goddess Dikē in denouncing perjurers and "'bribe eating' kings who render crooked judgments.'" Lincoln notes that in some versions of the *Theogony*, ancient editors used the synonymous verb *mythēsasthai* ("to speak, to tell"). The term *mythēsasthai* is shown to be an appropriately synonymous term with *gērusasthai* to denote "speaking truth" by its appearance in the last line of the proem of *Works and Days*, where Hesiod calls upon Zeus for justice in legal proceedings, pledging in turn to speak the truth thus revealed:

> Zeus of the lofty thunder, you who dwell in the highest palace,
> Hear me, you who see and perceive: Straighten out the judgments, according to justice!
> And I will tell [*mythēsasthai*] real things to Perses.[141]

Thus, the speech of "proclaiming," *mythos*, is the speech of divinely mandated justice and truth.

In Hesiodic texts, *mythos* is also the speech used by the powerful, as when, in a fable related in *Works and Days*, the hawk seizes the sweet-voiced nightingale. The helpless nightingale weeps at her fate, but the hawk forcefully:

> spoke this *mythos* to her:
> "Good lady, why do you screech? One who is far your better has you.
> [...]
> Senseless is he who wishes to pit himself against those who are more powerful:
> He deprives himself of victory and suffers pains in disgrace."

*Mythos* is also the term given to oath by Hesiod in the *Theogony*, as when Zeus asks the warlike Ouranids, to help him fight against their brothers, the Titans. Their pledge of support is termed a *mythos*.[142]

The term also appears often in Homeric texts, as in Book 2 of the *Iliad*, when Odysseus prevents the Greek soldiers from fleeing to their ships. Clubbing them with his scepter, he shouts: "Sit still and hearken to the *mythos* of others, who are mightier than you: You, who are unwarlike, helpless, and not to be counted on in battle or assembly." Of the 167 instances of the noun *mythos* or the verb *mytheomai* in the *Iliad* counted by scholar Richard Martin, Lincoln reports that 97% of those instances occur in a situation "in which a

---

[141] Lincoln, 4.
[142] Ibid, 12-13.

powerful male either gives orders or makes boasts." Lincoln concludes, "A *mythos* is an assertive discourse of power and authority that represents itself as something to be believed and obeyed."[143]

The distinction between the speech of *logos* and the speech of *mythos*, notes Lincoln, is clearly characterized in the writings of Hesiod and Homer. Both poets' works are representative of the qualities of an oral culture in which "poetry is society's chief archival medium, as well as its most authoritative discourse and prime instrument for cultural reproduction over the course of generations." In the *Theogony* Hesiod claims that he has received gifts—presumably for both "recounting" and "proclaiming"—from the Muses, "that transformed him from the near-bestial state of the shepherd into that of the poet, close to the gods." As a poet, Hesiod, like all poets, is given knowledge of past memory by the Muses, heirs to their mother, Mnemosyne. Too, he is given the laurel-scepter of "specially privileged kings, priests, seers, and poets," a symbol of the deities' favor and presence in the living speaker. The laurel scepter is, along with the poet's lyre, an attribute of Apollo, from whom the poet receives knowledge of the future. Thus, the divinely inspired poet sings of "things past and those yet to come." This gift of the authority to proclaim is the gift of *mythos*. Further, observes Lincoln, "like all gifts in a precapitalist economy," the gifts of the Muses represent part of a process, rather than an end product; and, the relationship between donor and recipient is reciprocal. The poet mentions the Muses by name, expresses his gratitude, and invokes their continued presence. This reciprocity suggests the recursive nature of recounting, the gift of *logos*.[144] The poet in early Greek tradition possesses both *mythos* and *logos*.

Later, however, the rise of writing in Greece, Lincoln observes, provided an opportunity for study and reexamination of texts experienced outside the realm of performance, stripped of music, feasting and conviviality. By the sixth century BCE, criticism of poetry and poets began to appear in the writings of the Pre-Socratics. Xenophanes complained that

---

[143] Lincoln, 17.
[144] Ibid, 24-25.

Homer and Hesiod attributed to the gods all
The shameful things that are blameworthy among humans:
Stealing, committing adultery, and deceiving each other.[145]

In his longest extant poem, Xenophanes describes the ideal symposium, celebrated in a ritually purified chamber amidst luxury and elegance, provided with a simple and majestic banquet. "[M]en of good cheer...hymn the god / With well-spoken *mythoi* and pure *logoi*." By "pure *logoi*," Xenophanes means tales of "noble deeds" as opposed to "treating battles of the Titans, Giants, / Or Centaurs." These latter Xenophanes terms "fabrications [*plasmata* of earlier times," telling of "blameworthy" things that should not be recounted of the gods, "behaviors that would undermine important institutions (marriage, the family, law, commerce, the *polis*)." Xenophanes distinguishes these things from *mythoi*, a term he reserves for stories drawn from human memory, unaided by direct divine inspiration, that are "moral in their content, reverent in their attitude, and socially beneficial in their consequences." Noble deeds are retold, "As memory and striving for excellence make them known" to the teller. The quality of excellence is here understood as a human factor, not a divine one.

Democritus coins the term *mythoplasteontes* ("myth fabricators"), which he uses to denounce those who try to pass off falsehoods as sacred truths (*mythoi*). Lincoln points out that this term connects two lexical domains with which we are familiar: the noun traditionally used for true, authoritative, and trustworthy accounts (*mythos*), and the verb used for artisanal creations in malleable, impermanent materials such as clay, plaster, and wax and also in words and ideas (*plassō*, "to mold, form, fabricate").

The term *plassō* also applies to forgeries, counterfeits and fictions in general. Democritus thus distinguishes between those "whose *mythos* is true," which he equates with uncrafted and unvarnished, as opposed to those "whose *logoi* are many," implying trickiness and seductive techniques.[146] Democritus thus conflates cunning intelligence—*mêtis*—with making—*poiēsis*—and damns their conjunction as untrustworthy.

---

[145] Ibid, 26.
[146] Ibid, 30.

Empedocles (c. 495-535 BCE) claims divine status not only for his *mythoi* but for himself as he addresses his audience on his theory of the transmigration of souls (that souls descend from an original divine state through various forms on a path toward increasingly elevated incarnations to regain their empyrean status, shedding their mortality): "Know these things clearly, having heard this *mythos* from a god." There is a large difference, Lincoln reminds us, between Hesiod's claim that his poetic authority comes to him through the *mythos* spoken by the Muses, and Empedocles' claim to be the divine authority speaking the *mythos*. By the time Protagoras states "Man is the measure of all things" (in the latter half of the fifth century BCE), the very existence of the gods is called into question. Human life is too brief to determine the truth in the matter of the divine, but is competent to judge its own truths.

This also calls into question the authority of the poet's traditional claim to divine inspiration and with it an exalted place amid human activities. The question, Lincoln asserts, now becomes, "Do poets speak *mythos*, *logos* or both, and what value ought to be attached to these categories?" Gorgias (said to be the pupil of Empedocles) calls poetry nothing more than *logos* with meter. It manipulates opinions by playing on the emotions and otherwise persuades by means of methods that have nothing to do with truth. Its enchanting wiles work on the soul as drugs work on the body. In a rhetorical exercise in which he defends Helen of Troy against the poets who maligned her unjustly, Gorgias argues that Helen was deceived by Paris's seductive *logos*, and thus deceived was powerless to resist his persuasion to abandon her husband. Helen is therefore deserving of forgiveness, not revilement. In this exercise, Gorgias not only skillfully undermines the language and authority of the poet, but in effect directly accuses Homer of "having gotten her story wrong"—which means he could have been wrong about everything else.[147] Gorgias also shows his own language, that of reasoned argument, to be superior in persuasive power. This is the essence of Sophism. Yet, at the same time, Gorgias takes what modern thought would characterize as a relativistic stance in granting to the telling of myths an important moral status:

---

[147] Lincoln, 32-33.

Tragedy inspires and proclaims. It is something wonderful for people to see and hear, and produces deception through its *mythoi* and the passions it arouses. Further, one who deceives in this fashion is more just than one who does not, while he who has been deceived is wiser than one who has not.[148]

After exposing the manipulation of the poet's *logos*, Gorgias nevertheless grants a greater moral value to the poet's *mythoi* as carriers of wisdom, which is gained through the arousal of the passions and not through the moral force of reason. Not only is the question of poetic authority left unresolved, but the distinctions between *mythos* and *logos* appear increasingly blurred.

In writings attributed either to Euripides or Critias (ca. 460-403 BCE), the author praises the stratagem of a man "shrewd and wise" who first thought of inventing the undying, all-knowing gods in order to put fear into the hearts of men who might otherwise contemplate evil in secret. "Recounting these *logoi*"—telling these invented stories of the gods—and "having hidden the truth with a false logos," says the author, that shrewd man:

affirmed that the gods dwell in that place where,
By saying this, he could most frighten people.
As a result, he knew the fears that exist for mortals
And the advantages of a troubled life.[149]

Laws that prevent disorder and violence work well in the agora, but the law is incapable of following those with evil in their hearts into the privacy of their homes. However, fear of the just retribution of the gods can and a well-crafted *logos* can arouse this fear. This audaciously cynical passage, Lincoln notes, is not only a myth about myth, but purports to be a true story that puts to the lie traditionally accepted stories, claiming them as fictions (*pseudei logoi*) "fabricated and propagated by the state for purposes of its own."[150] Moreover, the actions of the state in employing these false stories as a tool for regulating society are morally correct, for by doing so it imposes the good of public morality in places where laws cannot reach. Whether Critias was the author of the foregoing passage or not, his biography fits the totalitarian worldview expressed in the text. Critias led the

---

[148] Gorgias, Fragment B23, qtd. in Lincoln, 34.
[149] Lincoln, 35.
[150] Ibid, 35-36.

viciously repressive Spartan-backed "Tyranny of Thirty" that took
power in Athens after its defeat in the Peloponnesian War and carried
out a reign of terror that lasted nearly a year until it was overthrown
and Critias killed.[151] The above passage, says Lincoln, well states his
political views that "a small elite can and ought to impose moral
order on citizens, who are by nature weak, unruly, and given to secret
sins."[152] And further, that in order to carry out its ends, this elite is
justified in its means.

The thought of Critias, states Lincoln, leads directly to that of his
kinsman and fellow student, Plato.[153] Plato follows Xenophanes in
condemning certain traditional poetic themes. Tales of battles among
the gods incite civil strife. Tales of the hopeless underworld sap the
courage of soldiers. Like Gorgias, Plato[154] views poetry as a form of
*logos* appealingly enhanced by melody, rhythm and meter, but none
of these things make it true, which is to say verifiable through
analytic rigor. Moreover, poetry renders the populace lazy, seeking
the pleasure of images rather than seeking truth. Plato [in *Phaedrus,
Ion* and *Laws*] allows that inspired speech sometimes flows through
the poet, but its force renders him, temporarily, divinely mad, so that
he functions as a mere transmitter who himself adds nothing. Mere
poetic skill withers in the face of the power of this legitimate
madness.[155]

Plato categorizes *mythoi* as a form of *logos*, a category which
possesses some truth but is false on the whole and morally defective.
However, poetry has a place, and a dual purpose. It can speak with
certainty on topics which philosophical inquiry cannot, such as the
nature of the gods and the fate of the soul after death. And, following
the spirit of the author of the passage praising the shrewd man who
invented the gods, Plato envisions the use of *mythoi* to indoctrinate
segments of the populace—women, children, and the lower classes—
who are unable to follow the subtle philosophical arguments that
persuade rational men to propositions deemed necessary to the good
of the state. Lincoln concludes:

---

[151] Lincoln, 36.
[152] Ibid, 37.
[153] Ibid, 37.
[154] *Gorgias*, 502c.
[155] Lincoln, 38.

In the network of communicative relations envisioned by Plato, poets—who understood themselves to mediate between gods and humans—were significantly repositioned. The space he assigned to them is that which lies between the state and its lowliest subjects, where they craft *mythoi* at the direction of philosopher-kings, for mothers and nurses to pass on to their charges. And in this system, *mythoi* were not only revised but also radically revalorized.[156]

In Plato's ideal republic, myth finds its status reconfigured and reduced. It no longer emanates from the gods and heroes but becomes a property of the state, properly constrained into the well-ordered structure mandated by elite philosophers. Its ambiguous and unsettling aspects are discredited and deprived of voice.

*Mythos* has been shown to have changed its meaning. What then of *logos*? *Logos* as presented in the Platonic dialogues remains a difficult and much-discussed philosophical concept. Socrates uses the analogy of the sun, light, and vision to describe the relation of the unchanging ideal or "good" to *logos*. The sun, says Socrates, "is not vision, yet as being the cause thereof is beheld by vision itself." The good stands in a similar relation to reason, by means of which the good may be perceived. When illumined by reason, the mind apprehends truth and reality; but "when it inclines to that region which is mingled with darkness, the world of becoming and being passing away, it opines only and its edge is blunted, and it shifts its opinions hither and thither, and again seems as if it lacked reason."[157] In another dialogue, the sun metaphor reappears. One looks at the sun only indirectly as during an eclipse, never through mediation of the senses, lest, Socrates says, the soul be blinded. Rather, one must turn to the world of theories [*en logois*] "and use them in trying to discover the truth about things."[158] "We predicate 'to be' of many beautiful things and many good things, saying of them severally that they are, and so define them in our speech [*logōi*]."[159]

Yet, how does the speech of *logos* determine 'truth'? In Plato's *Phaedrus*, Sophocles has led Phaedrus to agree that the contention of rhetoricians in law courts of what is just and unjust relies on speech,

---

[156] Ibid, 42.

[157] Plato, *Republic*, 6:507, 508, *Collected Dialogues*.

[158] Plato, *Phaedo, Republic*, 99d-99e, *Collected Dialogues*.

[159] Plato, *Republic*, 7:507b, *Collected Dialogues*.

and asks, is it not true that, "He who possesses the art of doing this can make the same thing appear to the same people now just, now unjust, at will?"[160] Not only this, but if the speaker in the law court wishes to mislead someone else, he may do this through shifting his ground "little by little," so as to be more able to "pass undetected from so-and-so to its opposite" than he could do "in one bound." If this is so, then, "It follows that anyone who intends to mislead another, without being misled himself, must discern precisely the degree of resemblance and dissimilarity between this and that." With Phaedrus agreeing that this is essential, Socrates continues, "Then if he does not know the truth about a given thing, how is he going to discern the degree of resemblance between the unknown thing and other things?" Even he who intends to mislead can only do so with "knowledge of what the thing in question really is."[161] In other words, the truth of a thing must be known, ironically, even if one intends to deceive. "Thus," remarks John Sallis of this passage, "it appears that there can be no effective rhetoric independently of the knowledge of things."[162]

Some things we possess a prior shared vision of: "When someone utters the word 'iron' or 'silver,' we all have the same object before our minds, haven't we?" "But what about the words 'just' and 'good?'"[163] We disagree, we dispute with each other and ourselves. "This means that speaking is an essential means by which we are able to come closer to the things themselves in their truth, that it belongs to the means by which man is able to make the things themselves manifest."[164] If rhetoric has to do with "a kind of leading the soul by *logoi*," dialectic unfolds things into manifestness. How it does this, and what makes speech beautiful, proportionate and expressive of the good, is by first "gathering" and then "dividing." As an example Socrates uses the speeches on love that are the subject of conversation between him and Phaedrus at the beginning of their walk outside the walls of Athens on this particular day. Both began

---

[160] Plato, *Phaedrus*, 261c-d, *Collected Dialogues*.
[161] Plato, *Phaedrus*, 262a-b, *Collected Dialogues*.
[162] John Sallis, *Being and Logos: The Way of Platonic Dialogue*, 2nd. ed. (Atlantic Highlands: Humanities Press International, Inc., 1986), 169.
[163] Plato, *Phaedrus*, 263a, *Collected Dialogues*.
[164] Sallis, 170.

with the idea of "madness" in connection with love. Like a skilled butcher who divides the leg at the natural joints rather than clumsily hacking at it, the first speech proceeded from the main idea by dividing off "a part on the left, and continued to make divisions, never desisting until it discovered one particular part bearing the name of 'sinister' love, on which it very properly poured abuse." The second speech "conducted us to the forms of madness which lay on the right-hand side," until it discovered "the type of love that shared its name with the other but was divine" displaying it to our view and extolling it "as the source of the greatest goods that can befall us."[165] In this example, Sophocles himself is "collecting" these two examples of "division," and in these examples, "it becomes evident that perfected speech takes the form of collection and division, that is, of dialectic."[166] Socrates tells Phaedrus,

> Believe me, Phaedrus, I am myself a lover of these divisions and collections, that I may gain the power to speak and to think, and whenever I deem another man able to discern an objective unity and plurality, I follow "in his footsteps where he leadeth as a god." Furthermore—whether I am right or wrong in doing so, God alone knows—it is those that have this ability whom for the present I call dialecticians.[167]

*Legein,* the verb form of the word *logos,* means "both to say, to speak, and to lay in the sense of bringing things to lie together, collecting them, gathering them together."[168] Socrates' definition of dialectics contains both. *Logos* needs to be spoken to carry power and truth. Further, Sophocles implies that living speech occurs in the gathering together of dialogue, wherein we discover the truth of things through disputation.

Socrates remarks to Phaedrus that,

> The painter's products stand before us as though they were alive, but if you question them, they maintain a most majestic silence. It is the same with written words; they seem to talk to you as though they were intelligent, but if you ask them anything about what they say,

---

[165] Plato, *Phaedrus,* 266a-b, *Collected Dialogues.*
[166] Sallis, 170.
[167] Plato, *Phaedrus,* 266b, *Collected Dialogues.*
[168] Sallis, 7.

from a desire to be instructed, they just go on telling you just the same thing forever.

A text does not know how to address the right audience, nor can it prevent itself from addressing the wrong people. And, Sophocles continues playfully, "when it is ill-treated and unfairly abused it always needs its parents to come to its help, being unable to defend or help itself."[169] As a painting is a mere representation of reality, so a written speech, as Phaedrus responds, "may be fairly called a kind of image," not living speech.[170]

The problem remains how to determine whether a *logos* reflects the good and contains truth. Roochnik points out that by the time of Protagoras, who asserts that human beings make their own values, meanings, and purposes and that the human world is in effect *produced* by human, not godly, activity, this "truth" is no longer a matter that is determined by the gods and revealed through the *mythos* of heroes and poets. "Values finally become a matter only of which creator has the most power to impose his particular version of value on the rest of us.... It doesn't matter what values are created, only whether or not they can be successfully imposed."[171]

What are *mythos* and *logos*? The researches and arguments presented above are inconclusive, but point to fundamental issues of power and authority within the community. Bruce Lincoln's research demonstrates that *mythos* is originally a speech of the gods or inspired men and contains truth; *logos* is the persuasive, untruthful, twisted speech of women and weaklings. By Plato's time the meanings have essentially reversed: *mythos* is not to be trusted; *logos*, in Socrates' sense, walks with a "straight" gait, in the sunlight of reason. As *technê* shifts from its ancient centeredness within communal well-being to an evaluative abstraction which alienates imagery and imagination from making and restricts it to a quantifiable product as opposed to a qualitative process; so *logos* comes to represent a concept which is abstracted from the context of relativistic human flux.

---

[169] Plato, *Phaedrus*, 275d-e, *Collected Dialogues*.
[170] Plato, *Phaedrus*, 276a, *Collected Dialogues*.
[171] David Roochnik, *The Tragedy of Reason: Toward a Platonic Conception of Logos*. New York: Routledge, 1990, 96.

What then are the implications for the Hephaistean maker, who combines both straight-gaited *technê* and crooked-walking (and speaking) *mêtis*? This question will be carried into an examination of the Hephaistos myth in Chapters 3 and 4, with special regard to the mythic Hephaistos' relation to the authority and power of father and mother, together with implications within the *polis* of the double image of legitimacy versus bastardy for the ethical question of "straight" vs "crooked" speech, and its relation to "truth" in art and story, as well as to "ethics" in technology.

## *Plato and the Unreliability of the Imagination*

Plato's metaphysics places the original forms of "Being" in an immutable, timeless, and purely good realm of Ideas, in imitation of which the Demiurge, or original creator/craftsman described in *Timaeus*, created the material world. Human life exists in this transient and by definition imperfect plane of imitation, cut off from direct knowledge of the world of Ideas.[172] Not only are the human arts alone insufficient to produce the good of the *polis*, but the material world itself is only a copy created by the demiurge described by Plato in *Timaeus*. It is a mere imitation of the original Idea. Every subsequent act of creation is a copy of a copy, and humankind is a "race of imitators" (*ethnos mimētikon*), poor mimics who produce "no more than unreal artifacts"—such as the shadows in the famous analogy of the cave in Plato's *Republic*, which are merely "man-made images (*eidōla/phantasmata*)" and illusions (*eikasia*). This metaphysical condition requires a separation of reason and imagination. Inside the cave, the dark glass of imagination; outside, the light of reason. The fantasy, the image, the icon can never represent truth.[173] Only reason "can elevate the noblest part of the mind to a contemplation of the highest being."[174]

Richard Kearney provides a useful summary of Plato's retelling of the myth of Prometheus. In the *Protagoras*, "Plato explicitly associates the 'art of making' (*demiourgikē technē*) with the

---

[172] Richard Kearney, *The Wake of Imagination* (Minneapolis: University of Minnesota Press, 1988), 88.
[173] Ibid, 90-91.
[174] Plato, *Republic* 532c; Kearney, 91.

Promethean gift of stolen fire." With the gift of Prometheus, humanity, in Plato's words, "had a share in the portion of the gods." Humans were thereby able to transform themselves from the nature of animals to the nature of cultivated humans. Through the gift of the arts stolen from Hephaistos and Athena by Prometheus, humans discovered articulate speech and invented all the good things of material life from the matter of earth. However, they lacked an art which was in the exclusive keeping of Zeus and to which Prometheus had no access—the higher art of politics (*technê politikê*). Without the gifts of justice, moderation, and order that structure the *polis*, unguided humans quickly ran amok.[175] It was Zeus, not Prometheus, who granted the gift that rescued humanity from inevitable self-destruction, sending Hermes "to impart to men 'the quality of respect for others and a sense of justice, so as to bring order into our cities and create a bond of friendship and union.'"[176] Thus, Prometheus' gift to humanity of the arts of making (*demiourgikê technê*) could lead only to ruin. Kearney notes that "By thus polarizing the rational order of Zeus and the imaginative disorder of man," Plato separates the divine good from human evil.

Understanding the concept of the inferior value of the image—the gift of Prometheus—within the context of the divine order of the *polis*—the gift of Zeus—it becomes clear, suggests Kearney, exactly why the artist is excluded from the ideal *polis* Plato describes in the *Republic*. Plato indicts the image, its maker and human imagination on five points.

First, the human imagination is *ignorant* of the "ultimate nature of things." The artist's representation is a long way removed from truth. He is able to reproduce everything because he never penetrates beneath the superficial appearance of anything. For example, a painter can paint a portrait of a shoemaker or carpenter or any other craftsman without knowing anything about their crafts at all; yet if he is skillful enough, his portrait of a carpenter may, at a distance, deceive children or simple people into thinking it is a real carpenter.

The bed made by the carpenter is a copy of the original Idea of the bed, made by God. When the painter depicts the carpenter at his

---

[175] Kearney, 88-90.
[176] Plato, *Protagoras* 322c, qtd. in Kearney, 89.

trade, he "*simply* imitates what the other two *make*."[177] His work therefore is no more than a lie (*pseudos*). Plato shows that this also applies to the maker of tragedies, who as an imitator like the painter, is "in his nature three removes from the king and the truth, as are all other imitators."[178] When the argument is pressed, even the smith who makes the horse's bit in his forge cannot be said to know as much of its true nature as the horseman. Thus, Plato defines three categories of knowledge at subsequent removes from the truth of the Idea: (1) the user; (2) the maker; (3) the imitator.[179]

Second, the works of the artist are *nondidactic*. They teach nothing about reality and produce no useful results. They do not instrumentally influence meaningful human affairs and are therefore *useless*. Homer's poetry had no political impact comparable, for example, to Lycurgus reforming the constitution at Sparta. Homer never waged war, never commanded an army or invented a useful device, nor taught enthusiastic pupils a practical way of life.[180]

Third, the artistic imagination is *irrational*. To the extent that images appeal to "erotic and animal desires," they are equally removed from reason and rational calm—and thereby removed from the sense of truth and justice that define the higher qualities of rational human community and the good of the *polis*.[181]

Fourth, art is corrupting and *immoral*. It leads its audience into imaginative empathy with the faults it presents. Its allures cause vulnerable humankind to falter, lose control, succumb to laughter and foolishness. "Poetry has the same effect on us when it represents sex and anger, and the other desires and feelings of pleasure and pain.... It feeds them when they ought to be starved, and makes them control us when we ought...to control them."[182]

Fifth, art tends to *idolatry*. "The artist 'takes a mirror, and turns it round in every direction...thus rapidly making the sun and the heavenly bodies, the earth and even himself.'"[183] This mirror reflects

---

[177] Kearney, 90-92.

[178] Plato, *Republic*, 597e, *Collected Dialogues*.

[179] Plato, *Republic*, 601c, *Collected Dialogues*.

[180] Kearney, 92-93.

[181] Ibid, 93.

[182] Plato, *Republic*, 606d; Kearney, 94.

[183] Plato, *Republic* 596d, qtd. in Kearney, 94, italics mine.

only that made by man; not that made by the divine. Thus, Kearney observes, for Plato, "Imagination is idolatrous to the extent that it worships its own imitations instead of the divine original."[184] Thus, mythic images of the divine portrayed by poets such as Homer, Hesiod and Aeschylus are "conducive to blasphemy."[185] Further, Plato calls the image a "poor child of foster parents." "The mimetic image is an illegitimate son, who like the Stranger in Plato's *Theaetetus*, 'dares to lay unfilial hands on the paternal pronouncement' (*patrikoi logoi*)."

Here, Kearney resorts to Jacques Derrida's discussion of *logos*. Plato's model of *logos* entails the concept of "divine being as an original presence to itself. The mimetic image is a threat to this original presence, the dialogue of being with itself, for it constitutes a detour of representation—Derrida would say *écriture*—which claims 'to do without the Father of logos.'" The work of the artist competes with this self-identity, transgressing against the "Law of the Father."[186] The artist who styles himself as a human demiurge, by contrast, becomes the "Father's other." The distinguishing mark of all artistic or imaginative discourse is that "it cannot be assigned a fixed spot...sly, slippery and masked...a joker, a floating signifier, a wild card which puts play into play."[187]

Kearney continues, "It is precisely because the imagination introduces indeterminacy and ambivalence into discourse that it serves to deconstruct the paternal *logos* of self-identity," which seeks, ultimately, "to replace the original *presence* of divine being to itself."[188] It follows that this ambiguous play is to be regarded with suspicion. It must be controlled as a father controls the play of children by providing rules and stable conditions. Stable parental control in childhood, Plato counsels in the *Laws*, ensures peace and stability in adulthood, and maintains the good of the status quo.

Because Plato uses the imaginative mode in dialogues that condemn the imagination, a distinction must be understood between the legitimate functions of image as essentially a means to an end,

---

[184] Kearney, 94.
[185] Ibid, 98.
[186] Ibid, 95-96.
[187] Jacques Derrida, *Dissemination*, qtd. in Kearney, 96.
[188] Kearney, 96.

and its use by artists as an end in itself.[189] This remains problematic for Plato who must concede, in the *Sophist*, that pure thought occurs in mental images. Moreover, in the *Republic*, "imagination might be permitted a certain valid function in [Plato's] ideal state if it agreed to renounce its claim to artistic autonomy and submit to transcendental guidelines, 'producing hymns to the gods and paeans in praise of the good.'"[190] Too, dreams and ecstasies—forms of divine inspiration (*enthousiasmos*)—emanate from the gods, not the mind of man, and are to be distinguished from the manipulative tricks of the artist. Nevertheless, these true and corrective visions are rarely understandable to those who receive them, and must be interpreted by rational commentators. Finally, image is always subordinate to reason.[191]

## *Aristotle's Constructive Phantasia*

Aristotle (384-322 BCE), notes Kearney, shifts the argument concerning the relation of the imagination to truth or falsehood from a metaphysical to a psychological assessment. Far from being an illegitimate and troublesome tendency that is nevertheless difficult to completely expunge from the ideal *polis*, Aristotle views *phantasia* as a cognitive factor, mediating between the senses and the faculty of reason. Where Plato's image is external to the real, for Aristotle the mental representation (*phantasma*) stands between the outer world of experience and the inner world of knowledge. The imagination represents the world of experienced phenomena to the faculty of reason, which without the mediation of image would be empty of content. This imaginative representation constitutes a movement between mere sensation to cognitive potency resulting in the creation of knowledge and the ability to act.

Kearney discusses how, in the *Rhetoric*, Aristotle explains this movement. The imagination is moved by *desire*, specifically the desire of all humans to gain knowledge. Humans create imaginary images (*phantasma*) of anticipated evils such as pain or goods such as love. This gives imagination an ethical dimension, Kearney

---

[189] Kearney, 98-99.

[190] Plato, *Republic*, 607a; Kearney, 103.

[191] Kearney, 105.

explains, in that "Imagination is thus accredited a central role in the moral orientation of behaviour" through the creation of pictures that affect appetites and motivate choice. The imagination is also implicated in *time*. These mental images are of a temporal nature. They "recall our experience of the past and...anticipate our experience of the future."[192] Memory aids in imagining *"possible modes of experience,"* and thus in moving from sensation to higher cognition:

> The thinking soul apprehends the forms in images, and since it is by means of these images that it determines what is to be sought and what avoided, it moves beyond sensation when it is concerned with such images....Moreover, it is by means of the images or thoughts in the soul, which enable us to see (the future), that we calculate and deliberate about the relationship of things future to things present.[193]

The transformation from passive sensation to cognitive empowerment Aristotle recognizes as the "active intellect" (*nous poiētikos*). To further qualify, Aristotle distinguishes between two types of imagination:

> Where the sensible imagination refers exclusively to our empirical appetites, the rational imagination is capable of uniting and combining our empirical sensations in terms of a 'common sense,' which is in turn representable to reason. This 'synthetic' practice—which Aristotle terms *eidōlopoiountes*—is a unique property of the rational imagination.[194]

The *eidōlopoiountes*, "image-making," can be paraphrased as the *making* intellect, and is necessary to creation of actions and things in the world.

In light of the foregoing, it is not difficult to see how Aristotle is able to place a positive ethical value on the transformative power of the imagination. In the Aristotelian conception, the mimetic function of art, instead of separating humanity further from reality, instead affirms the universal meaning of life, giving human existence "a heightened sense of unity and coherence." It is a *mythos*, dealing in "essences rather than appearances" and thereby renders truth, not falsehood.[195]

---

[192] Kearney, 109-10.
[193] Aristotle, *De Anima* 3,7, 431b, qtd. in Kearney, 111.
[194] Kearney, 111.

According to Kearney, the European heritage of paradigms of the imagination has issued in two "excesses" situated uncomfortably at opposite ends of a continuum of meaning-making:

1. The premodern tendency to repress human creativity in the name of some immutable cause which jealously guards the copyright of 'original' meaning;

2. The modern tendency to overemphasize the sovereign role of the autonomous individual as sole source of meaning.[196]

Neither is tenable. One view is bleakly and hopelessly limiting; the other amounts, in the end, to a self-referential hall of mirrors, a narcissistic void.[197] Recovering the full spectrum of Hephaistean mythos may provide a way to re-image the imagination and its relation to effective action in the world.

This study seeks to regain the lineaments of a lost—or broken—archetype by recovering and re-making the myth of Hephaistos. This chapter examined terms that arose in the Greek literature between the period of Homer and Hesiod and that of Plato and Aristotle, terms which are relevant to the restoration of strands of Hephaistos's myth that have been forgotten, elided, or altered.

---

[195] Ibid, 106.
[196] Ibid, 33.
[197] Ibid, 188.

CHAPTER 3

# Greek Hephaistos

HEPHAISTOS, son of Hera, is the Greek god of fire and the smithy. Hephaistos is said to be the only Olympian who works. He is the maker of the bronze mansions and furniture of the Olympian gods, and of famous jewels and accoutrements of war used by gods and privileged mortals. He works not only in iron and noble metals but also in clay, the medium from which he fashions the body of Pandora.

He is the only Olympian god characterized as "ugly" and is the only who is lame, possessing a strong upper body but impaired legs or feet (*kullopodíon*, "crook-footed"), which cause him to walk with a circular, shuffling gait. Hephaistos is also the only Olympian Homer tells us, who has felt "mortal pain." He is the only Olympian male god who divorces. His sexual potency seems in doubt.

The visual attributes of Hephaistos are relatively few. When pictured, he is shown mature and bearded, very often wearing a *pileus* cap and a short robe appropriate to a worker in the forge. He often carries tongs or hammer. As noted above, in a few known representations his feet are twisted backward, though more often it is difficult to discern his impairment. His place associations are mostly abstractions rather than specific locales and include mountain/sky (Olympus), earth and the chthonic regions (the volcanoes thought to be his forges), and sea (his refuge with the sea-nymphs Thetis and Eurynome). One specific place association is with the island of Lemnos, located in the northern Aegean, close to Asia Minor, though relatively little is known of the origins or age of his cult there. In

Athens he shared with Athena the parentage of the Athenian people. His Athenian temple (449 BCE) is largely intact and is one of the largest of the surviving buildings of the classical period, and he was accorded an altar in the Erechtheion. He was honored at the Apaturia, the festival at which Athenian citizens celebrated their descent from a common ancestor, Erichthonios, the snake-tailed son of Hephaistos engendered on Gaia and foster-mothered by Athena. He was also honored in various festivals including a torch race conducted as part of the Panathenia. Aspects of Hephaistos' cult function in Athens remain unclear, but he was closely associated with the artisans of Athens who claimed the god as their patron.

This chapter assembles the Greek writings and traditions on Hephaistos and allied figures. No archetype manifests in isolation and indeed, Hephaistos appears in constellation with other deities in specific ways. Some of the questions this chapter will explore are: What is the significance of Hephaistos' parentage? What is the nature of his connections with goddess figures such as Hera, Thetis, Aphrodite and the Graces, Athena, Gaia, and Pandora? What is the nature of his connections with the gods, especially Zeus, Dionysos, Hermes, and Ares? What is the significance of art and craft for the ancient Greeks as expressed in Hephaistos' connection, and sometimes identification, with Prometheus and Daidalos? What clues can be found that affirm his connection with magic and his possible origins?

### Homer's Version: Four Scenes with Commentary

No more extensive composite portrait of Hephaistos appears in ancient literature than in Homer's *Iliad* and *Odyssey*, nor have any other sources had more influence on how the god has been remembered since Greek times.

In the *Iliad*, Hephaistos features in three scenes. In Book 1, Hephaistos appears as a regular family member present at the Olympian banquet table, where he acts as the peacemaker when a quarrel breaks out between Zeus and Hera. He calms Hera and reassures his brother and sister gods when the threat of paternal violence looms. The second and the most extended is the visit of Thetis to Hephaistos' workshop, which results in his making of a war shield for her son Achilles. The third Iliadic scene depicts Hephaistos' only direct involvement in the Trojan war, when nearly all the gods have joined the fray on one side or the other, and he is called to the field by Hera to cast his powerful, withering fire on the river god Xanthus who is threatening Achilles.

In the *Odyssey*, Hephaistos appears within a set of tales-within-a-tale told at the court of Alcinous by the blind bard Demodocus, whose skill arouses the tears of Odysseus and moves him to choose this moment to reveal his identity to the company and share his own story in turn. Here we are told of the magical net discussed in Chapter 2 as emblematic of Hephaistos' ensnaring *mêtis*.

Each of these four scenes unfolds a key element of the character of Hephaistos. It is this character—peacemaker, gentle clown, dutiful son with a streak of fiery anger appropriate as a response to the violence he has suffered through maternal and paternal rejection and cuckholding—which persists into his subsequent mythopoetic representations. Each scene also provides clues to deep ambiguities which will be revealed by a more detailed examination of the Homeric episodes, as well as through comparison to the more fragmentary Hephaistean traces in the writings of other classical authors and through evidence in Greek visual representations. These ambiguities will be explored in this chapter and amplified in Chapter 4 through comparison and contrast to related mythologems in other ancient sources and non-Greek traditions.

### Scene 1: At the Gods' Banquet

The precipitating action preceding this scene begins with the appearance of Thetis, who breaks "from a cresting wave at first light" and soars to where Zeus sits enthroned on the uppermost crown of Olympus, "gazing down at the world."[198] Thetis grasps his knees with her left hand and cups his chin with her right and elicits his promise that he will grant the Trojans "victory after victory," frustrating the Achaeans until they are forced to beg her son Achilles to fight with them, covering him with honors to atone for Agamemnon's arrogant scorning of him. (This significant moment is depicted in Jean Auguste Dominique Ingres' monumental 1811 painting, *Jupiter and*

---

[198] Homer, *Iliad*, trans. Fagles, 1:590-95.

*Thetis.*[199]) Zeus (Jupiter) bends himself to Thetis's importuning, knowing that it will delay his own plans and also anger Hera, who vengefully champions the Achaeans as punishment for Paris's failure to judge her most beautiful of the goddesses. Indeed, Hera has seen Thetis with Zeus and reproaches him for scheming against her. She badgers him until he threatens her, thundering,

"Obey my orders
for fear the gods, however many Olympus holds
are powerless to protect you when I come
to throttle you with my irresistible hands."

Hera is terrified and "throughout the halls of Zeus the gods of heaven quaked with fear."[200] Hephaistos alone rises from the gods' banquet table, gently warning Hera that Zeus is far too strong, that she must go back to him with "soft, winning words" and persuade him from his mood. He places a cup into Hera's hands and urges her:

"Patience, mother!
Grieved as you are, bear up, or dear as you are,
I have to see you beaten right before my eyes.
It's hard to fight the Olympian strength for strength.
You remember the last time I rushed to your defense?
He seized my foot, he hurled me off the tremendous threshold
and all day long I dropped, I was dead weight and then,
when the sun went down, down I plunged in Lemnos,
little breath left in me. But the mortals there
soon nursed a fallen immortal back to life."

Hera, calmed and reassured, smilingly takes the cup from her child's hands. Fear turning to relief, "uncontrollable laughter broke from the happy gods as they watched the god of fire breathing hard and bustling through the halls," pouring out nectar right and left for all present. They feast until the end of day and retire, each "to rest in his own house, the splendid high halls Hephaistos built for each / with all his craft and cunning, the famous crippled Smith."[201]

This passage introduces many facts about Hephaistos: he has come to his "dear" mother Hera's defense in the past—we don't

[199] Jean Auguste Dominique Ingres, *Jupiter and Thetis*, 1811, oil on canvas, 345 cm. x 257 cm., Aix en-Provence, Musée Granet.
[200] Homer, *Iliad*, trans. Fagles, 1:680-667.
[201] Ibid, 1:706-731.

know exactly how, whether through words or deeds, openly or covertly—and learned to rue the day Zeus angrily seized him by the foot and hurled him so that he fell, for a full day, to land broken on the earthly island of Lemnos. As a result, he has a relationship with the island's people; not one arising from exercise of godly power over the mortals there, but from his vulnerability. Having come closer to mortal death than any of his brother and sister Olympians, he has reason to recognize the immutable power of Zeus's rage when aroused. Yet, he is the only one among the gods who rises to their mutual defense, which he does by taking the fire out of the situation. He speaks to Hera as a child interested in protecting his mother, but none of the others, quaking with fear, dares to intervene in fear of an outburst from Zeus which could send their banquet "crashing," yet again, and "blast" *all* of them from their seats.[202]

Hephaistos' speech can be interpreted as being a public criticism of Zeus, as it recalls an act of violence the result of which was that Hephaistos, his son, suffered a grievous wounding at his hands. Yet he couches his speech and behavior as he passes Hera a cup (undoubtedly of his own making) in such a way that he both takes the edge off Hera's anger and fear and causes all the gods to laugh as they watch him as he limps to fill their cups.[203] For all we know at this point in Homer's telling, he limps as a direct result of Zeus's violence. Hephaistos has just delivered a double message, drawing attention to his woundedness and then away from it, by making light of that very woundedness. It is a matter of *techné*, the province of the only Olympian god who works, master not only of the *techné* of building the gods' mansions; but also of the *techné* of calming power and aggression with seductive *logoi*.

This would not entirely agree with Kenneth Atchity's view of Hephaistos as embodying "Homer's poetic consciousness of the governing order of Zeus." Atchity asserts that the smith-god and the poet are allied in purpose, each "in his own technical way" [i.e., through his *techné*] is an agent of order, "bringing the potential of Zeus' will into realization through their craftsmanship."[204] Homer's

---

[202] Homer, *Iliad*, trans. Fagles, 1:698-700.
[203] Ibid, 1:721-723.
[204] Kenneth John Atchity, *Homer's Iliad: The Shield of Memory* (Carbondale: Southern Illinois University Press, 1978), 134.

*ekphrasis* on the Shield is evidence that Homer the poet does make use of Hephaistos the smith-god as a representative of the ordered Zeusian universe and that Hephaistos as maker parallels Homer as maker. However, the indications of the god's *mêtis* and his skill with *logoi*, suggest a more hidden, volatile, and feminine-connected skew to the picture than Atchity, and indeed perhaps Homer envisioned. This is substantiated by other traditions that will be examined in the following chapter.

Following his theme of the ordered Zeusian universe, Atchity states that:

> Authority, the bulwark of peace, requires the exercise of the art of peace; courtesy and authority must be accepted lest all the gods suffer. Hephaistos himself not only counsels, but also exemplifies in his own speech, courteous propriety; he couches his suggestion to Hera in a diplomatic nicety.[205]

Hephaistos is a peacemaker whose speech preserves the authority of Zeus. Zeus, however, does not here speak of peace, but instead threatens violence, a threat that is allowed to hang in the offing. This theme is taken up centuries later by Aeschylus in *Prometheus Bound*. Aeschylus paints Zeus as the tyrant, to whom Prometheus is reunited in the end. But Hephaistos is not portrayed by Aeschylus as an unquestioning bulwark of the paternal authority of Zeus. Hephaistos duly carries out the task Zeus assigns him to chain Prometheus, but only after voicing his reluctance to do so and being prodded by Kratos ("Strength") to get on with it.

## Scene 2: In the Workshop of Hephaistos

The second Homeric scene involving Hephaistos reveals more about this god's indirect exercise of influence and power, perhaps in harmony with the larger designs of Zeus, but also perhaps not without the power to obliquely affect the outcome dictated by the ruler god.

A flight of Thetis again ushers in the scene. This time she comes from the battlefield of Troy (where Achilles is mourning the loss of his comrade Patroclus to whom Achilles had given his armor) to the workshop of Hephaistos on Olympus. She asks him to forge new armor for her half-divine, half-mortal son. Until now Achilles,

---

[205] Ibid, 139-40.

scorned and gravely insulted by Agamemnon, has held himself from the fight, but he will soon enter the field, clad in the armor and bearing the shield made for him by the god.

In this extended scene in Book 18 of the *Iliad*, Hephaistos is revealed as maker. First, Homer presents to his reader, through Thetis's eyes: "Hephaistos' house, / indestructible, bright as stars, shining among the gods, / built of bronze by the crippled Smith with his own hands." He is shown at work:

> There she found him, sweating, wheeling round his bellows,
> pressing the work on twenty three-legged cauldrons,
> an array to ring the walls inside his mansion.
> He'd bolted golden wheels to the legs of each
> so all on their own speed, at a nod from him,
> they could roll to halls where the gods convene
> then roll right home again—a marvel to behold.
> But not quite finished yet…
> the god still had to attach the inlaid handles.
> These he was just fitting, beating in the rivets.

At the beginning of the scene the laboring god is unaware of being seen, underscoring how normal his deep engagement in his work is. Homer describes Hephaistos as "an immense hulk" hobbling along on shrunken legs, moving nimbly. He has been working without a shirt and is sweaty and shaggy. It is his wife, Charis, "lithe and lovely," who sees Thetis and welcomes her inside the mansion which is also a workshop, seating her on a "handsome, well-wrought chair, / studded with silver…." That Charis wears a "glittering headdress," appropriate to her loveliness, suggests that this too is of Hephaistos' making, perhaps one of the elegant objects (*daedala*) formed in the "vaulted cave" under the ocean.

In the midst of such aesthetic magnificence, would one wonder that the Smith, crippled though he might be, possesses a beautiful wife whose name indicates she is one of the Graces? Charis offers gracious hospitality to Thetis and familiarly calls Hephaistos to welcome the guest: "Hephaistos, come— / look who's here!"

Hephaistos greets Thetis warmly and immediately states, perhaps for the benefit of Charis and certainly for the benefit of the reader, who she is in relation to himself:

> Thetis saved my life
> when the mortal pain came on me after my great fall,

thanks to my mother's will, that brazen bitch,
she wanted to hide me—because I was a cripple.
What shattering anguish I'd have suffered then
if Thetis had not taken me to her breast, Eurynome too,
the daughter of Ocean's stream that runs around the world.
Nine years I lived with both, forging bronze by the trove,
elegant brooches, whorled pins, necklaces, chokers, chains—
[*daedala*]
there in the vaulted cave—and round us the Ocean's currents
swirled in a foaming, roaring rush that never died.
And no one knew. Not a single god or mortal,
only Thetis and Eurynome knew—they saved me.[206]

Homer describes the shield itself in the famous *ekphrasis*, and he also depicts Hephaistos commencing to work in his workshop. He speaks to his twenty wheeled bellows, commanding them to work. Like his twenty cauldrons, they are animate. They blow on the crucibles, gauging their breath at the god's command, from heavy blast to the merest breath for the light work. He flings precious metals into the fire: bronze, tin, gold, silver. Heaving the anvil onto its block, he grips hammer in one hand and tongs in the other, his most commonly recognizable attributes for centuries thereafter.

Between the scene in Book 1 and the first 40 lines of the scene in Book 18, Homer has introduced at least two seeming contradictions, raising questions which the poet's text does not answer. First, has Hephaistos fallen from Olympus twice—once at the will of Hera, because he was already crippled at birth, and landing in the sea; and once thrown by angry Zeus, landing broken on Lemnos after falling for a full day? How many times has Hephaistos landed at the mercy of saviors, both mortal and godly? How many times has he felt "the mortal pain?" Second, Homer has introduced, for the second time in regard to Hephaistos, a scene of family and filial intimacy and speech that is both frank and persuasive, but this time in connection with his foster-mother Thetis instead of his "dear" mother Hera, whom he now terms a "bitch" within the third sentence he speaks upon Thetis' entrance. His plain speech follows immediately upon the impression Homer has created of Hephaistos' well-appointed home and married life with his lovely and hospitable wife—and if she is perhaps shocked by her husband's candor, Homer gives no indication.

---

[206] Homer, *Iliad*, trans. Fagles, 18:431-473.

### Scene 3: Firing on the River Xanthus

The third Iliadic scene depicting Hephaistos occurs in Book 21. Achilles, enraged at the death of Patroclus, has arrayed himself with the armor made for him by the god, and has entered the Trojan field with a vengeance, mercilessly hacking and slashing at the Trojans, literally piling up the bodies of those he has killed. The surge of battle backs into the ford of the River Xanthus, where corpses soon choke its very flow. The river god Xanthus appears in human form, crying out to Achilles to stop and if he cannot, to confine his butchery to the plain.[207] When Achilles refuses, Xanthus calls out to Apollo, challenging him to honor the commands of Zeus to stand by the Trojans. Hearing this, Achilles charges into the river and challenges the god himself. Xanthus, enraged in turn, rises in a flood. He threatens to drown Achilles in his massive waves, using the very weight of the armor made by Hephaistos to bury his body so deeply in mud and stones that the Achaeans will never be able to collect his broken and scattered bones. Alarmed at Achilles' distress, Hera summons Hephaistos:

> "To arms, my child—god of the crooked legs [*kullopodíon*]!
> You are the one we'd thought a worthy match
> for the whirling river Xanthus!
> Quick, rescue Achilles! Explode in a burst of fire [*piphauskeo de phloga pollen*]!
> I'll drive the West and South Winds with clouds
> and sweep in from the open seas a tearing gale to sear
> the Trojan bodies and gear and spread your lethal flames [*phlegma kakon*]!
> And you, you make for the Xanthus banks and burn the trees,
> hurl the stream itself into conflagration—not for a moment
> let him turn you back with his winning words or threats.
> Never abate your fury! Not till I let loose my shout—
> then halt your withering fire!"

Instantly, Hephaistos responds, launching his "grim inhuman blaze." His fire scorches the whole plain, burning the corpses of men Achilles has slaughtered and causing the river to shrink back into its banks. Turning his blaze toward the river itself, Hephaistos immolates the trees and plants of the river's banks, as Hera has

---

[207] Homer, *Iliad*, trans. Fagles, 21:237-250.

charged. Homer enumerates them: "the elms burned, the willows and tamarisks burned, and the lotus burned and the galingale and reeds and rushes." The river god Xanthus begins to gasp, calling on Hephaistos to stop his fire. Hephaistos does not. The river boils and stops flowing altogether. He cries out to Hera that he will never again support the Trojans, even to prevent the burning of Troy itself, if only Hera will call off Hephaistos. And hearing the river's capitulation, she does: "Hephaistos, stop! Stop, my glorious blazing boy! / It's not right to batter another deathless god, / not for the sake of these *mortals*."[208] Without a word, Hephaistos stops and the river resumes its flow.

This scene reveals much about the quality of Hephaistean fire for Homer. The fire Hephaistos unleashes on the river god is described as thorough, impersonal, and merciless. Note that these qualities are not attributed to Hephaistos personally, although in other connections, the sacrificial flame and the fire in the hearth are called "Hephaistos"—Hephaistos *is* the fire.[209] Though Robert Fagles, translator of the *Iliad* version I am using here, calls the blaze kindled by Hephaistos "grim and inhuman," Homer places in the mouth of Hera the term *kakon*, meaning "bad" or "evil" to describe the flames she wishes Hephaistos to muster. The Greek words for the action Hephaistos takes in response to Hera's command to aid her in stopping Xanthus are *titusketo thespiades pur*, "prepares god-kindled fire."[210] The epithet *thespiades* ("god-kindled") is not restricted in Homeric descriptions of fire to Hephaistos' creation, but also applies to the fiery deeds and characteristics of the epic's heroes. *Thespiades* also describes the fire Hector will cast on the Argive ships, an event that will mark the beginning of Zeus's intervention in answer to Thetis' prayer that the Achaeans have suffered enough at the hands of the Trojans to be ready to beg Achilles to come into the fight. The metaphor of fire plays through a sustained passage describing Hector

---

[208] Ibid, 21:377-432.

[209] Lewis Richard Farnell, *The Cults of the Greek States*, Vol. 5 (Chicago: Aegean, 1971), 374.

[210] Homer, *Iliad*, 21:342, *Homeri Opera*, Oxford University Press, 1920, Perseus Digital Library Project, ed. Gregory R. Crane, Tufts University, accessed September 12, 2004,
http://www.perseus.tufts.edu/cgi-bin/ptext?doc=Perseus%3Atext%3A1999.01.0133.

as godly, rampaging like Ares or a "flash fire" raging down a hillside. Its sense is that Zeus is casting on Hector a divine aura of glory, for this hero, whom Zeus wishes to honor, will soon be dead, sacrificed to the ending that Zeus will work on this great story. (Indeed, the description of Hector's burial occurs in the last line of the *Iliad*). Though delivered through mortal hands, and metaphorically illuminating a mortal man, this fire is god-kindled. Fiery Hector is also compared to waves and winds.[211]

The three elements of raging fire, mighty waves, and blasting winds also come together in the passage describing Hephaistos' encounter with Xanthus. The river's mighty waves are sizzled by Hephaistos' fire, propelled in a "torrid blast" that might be likened to the blowing of forge-bellows. Hera has also called upon Zephyrus and Notio, the West and South winds, to help propel the *phlegma kakon* ("evil flames").[212] At the end of Book 20, the enraged Achilles, about to launch his supernaturally massive single-handed slaughter amid the Trojans, is also described as blazing like an "inhuman fire," and a "huge fireball" swirled by the wind.[213] And, when the pyre of Patroclus fails to ignite, Achilles prays to Zephyrus and Boreas (the North Wind). Iris transmits the prayer and in response, the Winds whip up the seas and blast the earth, striking the pyre into a towering "god-kindled" blaze that lasts all night.[214] The final Homeric instance of *thespiades pur*, "god-kindled fire," occurs in the *Odyssey*, when Eidothea comes to the aid of Odysseus, warning him of the many forms her father The Old Man of the Sea (Proteus) will take, including "superhuman fire."[215] Thus fire, in addition to water, is an element associated with the shape-shifting possessors of *metîs*. It is only after Odysseus holds on to Proteus—exerting superior *metîs*—that he learns from the god what his fate is.

As soon as the river is subdued, a series of donnybrooks ensues between the gods ranged on opposing sides in support of the Trojans or the Achaeans. Ares challenges and is routed by Athena (who downs him with an enormous boulder). He is dragged ignominiously

---

[211] Homer, *Iliad*, 15:595-630, *Homeri Opera*.
[212] Ibid, 21:335-40a.
[213] Homer, *Odyssey*, trans. Fagles, 20:554-557
[214] Homer, *Iliad*, 23:210, *Homeri Opera*.
[215] Homer, *Odyssey*, trans. Fagles, 4:470.

from the field by Aphrodite (his divine body, inert, covers seven acres). Goaded on by Hera, Athena beats up Aphrodite as well, exulting. Poseidon then challenges Apollo, who demurs: "God of the earthquake—you'd think me hardly sane if I fought with you for the sake of wretched mortals." But his sister Artemis taunts him as spineless. Apollo refuses to reply, but Hera, hearing the remark, boxes Artemis' ears for her. "How do you have the gall, you shameless bitch" she spits, "to stand and fight me here?" Artemis runs away in tears to sit sobbing in Zeus's lap and to tell on Hera and be comforted. As Artemis' mother Leto collects the bow and quiver Hera has ripped from Artemis, Hermes, who has no intention of mixing it up with the rest, and has doubtless been enjoying the spectacle, assures her, "Nothing to fear, / I'd never fight with you, Leto.... / No, go boast to your heart's content and tell the gods you triumphed over me with your superhuman power."[216] Hephaistos, who has dispassionately and precisely fulfilled his mother's request without uttering a word, has long departed.

It is no wonder that the god's fire should be associated with the weapons of mass destruction created in modern times and unleashed impersonally, inhumanly, from great distance. But what this god's attitude is toward its unleashing is not described. How does the image of Hephaistos the peacemaker sit with the image of Hephaistos the unleasher of implacable fire? Uneasily, at best. However, the mythological image of the blacksmith, illustrated in many mythologies, contains just such seeming contradictions.

Atchity asserts that in the *Iliad*, "Hephaistos represents Homer's ambition to present a model for orderly behavior." Indeed, "[Hephaistos'] mind is wholly commanded by Zeus' new Olympian order." Further,

> Although Hephaistos seems to harbor no grudge against his mother, certainly his reference to her original unkind treatment of him may be taken as a qualification of his subservience to Hera; he will serve her faithfully only so long as her wishes do not contradict those of Zeus, whom Hephaistos fears and respects since the failure of the rebellion.[217]

---

[216] Homer, *Odyssey*, trans. Fagles, 21:442-572.
[217] Atchity, 137.

Whether and how much Hephaistos both "respects" and "fears" Zeus is a matter of interpretation, and will be further considered in this chapter in terms of Greek mythology and in the next through the lens of other Hephaistean mythic associations. It is clear that Homer has represented the god as in support of Zeus's universal order, and it can also be suggested, as Atchity does, "Smith-god and poet are closely allied," as evidenced in the *ekphrasis* on the Shield, which depicts the whole human world that is also the subject of the poet's wide ambition. Yet whether "both are agents of order, bringing the potential of Zeus' will into realization through their craftsmanship" deserves further discussion. As Atchity observes, Homer has subordinated the archaic chthonic elements of Hephaistos' myth into "the general impression of orderliness." Yet within this outward orderliness an element of unease remains. And, as will be seen below in the discussion of other mythemes, the order of Zeus is not absolute, and the craft of Hephaistos has many possible meanings.

### Scene 4: Demodocus' Tale of the Golden Net

In Book 8 of the *Odyssey*, the tale-within-a-tale of Hephaistos' response to the adultery of Ares and Aphrodite serves a different function in Homer's narrative than the tales of the gods' actions and interrelationships in the *Iliad*.

Odysseus, his identity as yet undisclosed, has arrived at Phaecia and been welcomed at the court of King Alcinous. The blind bard Demodocus sings three stories during the festivities recounted by Homer; the story of the golden net is their centerpiece. It strikes a different chord from the first of Demodocus' tales, "The Strife Between Odysseus and Achilles, Peleus' Son," that draws silent tears from Odysseus; and the third, requested by Odysseus, the tale of the Trojan Horse, after which Odysseus reveals his identity and tells his own tale.

The golden net crafted by Hephaistos to snare Ares and Aphrodite in their adulterous embrace in his own bed was discussed in Chapter 2 as symbolic of Hephaistos' *mêtis* as well as of his *technê*. Once again the limping god draws laughter from the Olympian assembly, whom he has summoned to witness his imagined triumph (the goddesses keeping to their own mansions out of modesty). This time, however, Hephaistos is not making peace but instead demanding witness and recompense. He is not calm, but rageful and indignant:

"Father Zeus look here—
the rest of you happy gods who live forever—
here is a sight to make you laugh, revolt you too!
Just because I am crippled, Zeus's daughter Aphrodite
will always spurn me and love that devastating Ares,
just because of his stunning looks and racer's legs
while I am a weakling, lame from birth, and who's to blame?
Both my parents—who else? If only they'd never bred me!
Just look at the two lovers…crawled inside my bed,
Locked in each other's arms—the sight makes me burn!"

Hephaistos makes a complaint against Aphrodite and an accusation against his parents that implicates them in his lame condition. The gods laughingly congratulate him: "Look how limping Hephaistos conquers War, / the quickest of all the gods who rule Olympus!" The cripple wins by craft." "The adulterer, he will pay the price!" Though Hephaistos has demonstrated his *metîs*, not only by ensnaring Ares and Aphrodite, but also through his feigning a trip to Lemnos—riding away on his donkey so that Ares will enter the trap unawares—his tone is not one of irony, and his demand is for divorce: "my cunning chains will bind them fast / till our Father pays my bride-gifts back in full, / all I handed him for that shameless bitch his daughter…." Poseidon urges him to release Ares, but Hephaistos refuses until Poseidon offers to guarantee Ares' debt for adultery. Once released, Ares escapes to Thrace and Aphrodite to Paphos, where the Graces attend to her [218]

This scene, so different from Homer's Iliadic depictions of Hephaistos, introduces new themes into the Homeric-Hephaistean mix. In it, Hephaistos is depicted as both jealous and rageful, as well as calculatedly vengeful. His vengeance by no means exceeds in degree the injury of the indignity committed against him; given the traditional punishments exacted for adultery in patriarchal cultures, the tale is clearly a burlesque. Indeed it is difficult to tell whether the gods laugh harder at the consternation of Ares and Aphrodite, or at Hephaistos' willingness to publicly expose his cuckholding—which Hermes would be quite happy to imitate, even if the penalty were to

---

[218] Homer, *Odyssey*, trans. Fagles, 8:347-410.

be ensnared in *flagrante delicto* (literally, "while the crime is blazing") with golden Aphrodite. It is largely to this scene (and to one additional mytheme, Athena's rejection of his sexual advances, to be discussed below) that the use modern writers have made of Hephaistos as a sexual bungler can be traced.

In sum, this scene adds to the questions raised about Hephaistos' origin. It is in this scene that he calls Zeus his "father," not in the generalized way that the patriarch of Olympus can be said to be the "father" of all the gods (even though he is brother to many) but in the specific sense of Zeus and Hera as father and mother who are biologically to blame for his crippled condition. The words Homer puts into his mouth are a whining complaint at his pitiable condition as an unattractive weakling whose honor is vulnerable to the wiles of a handsome, straight-limbed adulterer. Taken together with the complaint he makes against Hera when Thetis comes to his workshop in the *Iliad*, the psychological dimension of Hephaistos' wounding is compounded in this story. Bred and born defective, he has been twice thrown from Olympus, magnifying the stigma of lameness through rejection and punishment.

Atchity avers that in the first Iliadic scene in which Hephaistos appears, the Olympian banquet mentioned above, the opening words of Hephaistos' speech "express the consciousness of the social repercussions of marital disharmony, of antagonism between two individuals whose concord is the necessary basis for social order."[219] If this is so, and Hephaistos represents Zeusian order as Atchity insists, how shall we explain the disorder of Hephaistos' cuckholding and his response to it? Atchity insists that the "selfishness" of Aphrodite is disorderly, and Hephaistos is more properly mated with Charis, "the most nondescript, the most neutral, of the three [Graces]."[220] If so, the laughter drawn by Hephaistos would be a clownish counterpoint to the gravity of Zeus, and would be entirely contained within the patriarchal paradigm. But is this all there is to Hephaistos? Whose child is Hephaistos? What is the nature of his *technê*? What archetypal role does he play?

---

[219] Atchity, 139.
[220] Ibid, 136.

## *Alternate Texts, Additional Mythemes*

Although the Homeric version is the most elaborated accounting of Hephaistos' story, enticing and perhaps contradictory mythemes lurk in written fragments and in references to texts now lost, and in vase paintings.

### Hera's Anger

In Hesiod's Theogony Hephaistos' conception is plainly parthenogenic. Hera births Hephaistos in anger at Zeus. Why she is angry is not clearly stated, but this story appears after an account of the birthing of Athena from Zeus's head, implying this to be the cause of her anger. The same mytheme appears in a Hesiodic fragment, but this time the birth of Hephaistos occurs before instead of after the birth of Athena.[221] The nature of the *eris*, "quarrel," mentioned by Hesiod in both instances as the motivation for the deed is unspecified in the case of the fragment.

The "Homeric Hymn to Pythian Apollo" tells a fuller version of the story of Hera's anger at Zeus:

> Without me, he has given birth
> to bright-eyed Athena,
> who stands out from all the blessed gods.
> But my own boy, Hephaistos,
> the one I myself gave birth to,
> was weak among all the gods,
> and his foot was shriveled,
> why it was a disgrace to me,
> a shame in heaven,
> so I took him in my own hands
> and threw him out and he fell
> into the deep sea.
> The daughter of Nereus, Thetis,
> with her silver feet,
> took him and brought him up
> with her sisters.
> I wish she would have done us blessed gods
> some other favor!

---

[221] Gantz, 57.

Hera rails on, suggesting that Zeus has threatened her role as wife and queen: how dare Zeus give birth to Athena, alone, without her? "Wouldn't I have given birth for you?" Then, a threat: Zeus had better watch out lest she plan trouble for him. And, a plan: to bear a son "who will stand out among the immortal gods" —but this son will not come from Zeus. Hephaistos, a "disgrace," has been discarded and she wishes he had remained so.

To get the child she wishes who would threaten Zeus, Hera strikes the ground and prays to "Earth and Heaven, and the Titans, from whom we get both men and gods," to give her a child who is stronger than Zeus himself. She sees, happily, that Earth—Gaia—is "moved." Hera's conception will be assured by the very power of Earth and will echo Gaia's fecundating might. For a year Hera shuns Zeus's bed, won't even sit in her throne "giving him good advice." Instead, she remains in her temples, enjoying the sacrifices of the many people who pray to her.

> But when the months and days
> were finished, and the seasons
> came and went with the turning year,
> she bore something
> that didn't resemble the gods,
> or humans, at all: she bore
> the dreaded, the cruel, Typhaon,
> a sorrow for mankind.
> Immediately the lady Hera,
> With her cow-eyes, took it
> And gave it to her (the she-dragon),
> Bringing one wicked thing to another.[222]

This second foster-mother of a discarded son of Hera, the monstrous she-dragon, is later slain by Apollo. The mytheme of birth without the participation of a father is clear. So is the association of Hera's birthgiving with *eris* and with shame, resulting in deformity and the mother's acts of violent abandonment.

Marie Delcourt comments that Hera gives birth to Hephaistos in the same fashion, i.e., parthenogenically, as Gaia gives rise to Ouranos and Pontos, the first generation of gods.[223] Indeed, monsters

[222] "The Hymn to Pythian Apollo," *The Homeric Hymns*, trans. Charles Boer (Woodstock: Spring), 1979, 168-71.
[223] Marie Delcourt, *Hephaistos ou la lègende du Magicien* (Paris: Les Belles Lettres,

are generally born out of the Earth. In the case of the conception of Typhaon, Hera slaps the ground, and the earth, "source of life" *moves*. In the case of both monstrous births, that of the giant serpent and that of deformed Hephaistos, Delcourt sees evidence of a "telluric" Hera, a tradition little known but hinted at by Hesiod.[224]

In one version of the parentage of Hephaistos he is indeed the child of both Zeus and Hera. Hera and Zeus, who are both children of Kronos and Rheia and thus sister and brother, are betrothed by Okeanus and Tethys after Kronos has been banished to Tartarus. Soon after, Hera gives birth to Hephaistos. To save face and cover the shame of lying with Zeus in secret before marriage—they have met on the island of Samos—Hera claims the conception occurred without intercourse.[225] Delcourt remarks that this version of the tale sounds like "bourgeois moralizing" except that it also recalls a tradition attested to by Callimachus, as follows: on Samos (along with Argos, one of Hera's two most important cult-centers), the name *apulia* was given to a night when a young woman's fiancé slept in a separate chamber in her father's house, where he was joined by young girl with the same mother and father. Meanwhile, a young boy spent the night beside the fiancée in the house of the prospective son-in-law. Delcourt sees in this rite involving a symbolic spouse a souvenir of Hera's status as pre-Hellenic Great Goddess associated with an adolescent consort or son-spouse.[226] As will be seen in the next chapter, Hephaistos is often closely connected with initiation mysteries associated with the Great Goddess.

The *Iliad* provides a great deal of information about the ambivalent relationship between Hera and Hephaistos. Several ancient authors mention the story of the imprisoning throne sent by Hephaistos to Hera, which does not appear in the Homeric epics. The story of the binding of Hera appears in the writings of Alkaios (sixth to fifth centuries BCE) and others. A much later account (fourth century CE ) adds a significant detail concerning Dionysos: Hephaistos "pointedly" presents Dionysos to Hera as her "benefactor," who had succeeded in bringing the unwilling

1957), 32.
[224] Delcourt, 33.
[225] Gantz, 57.
[226] Delcourt, 34-38.

Hephaistos to Olympus to free his mother. Moreover, a relieved Hera is then persuaded to convince the other Olympians to admit Dionysos as one of their number.[227] (This would be yet another example of the craft of Hephaistos in using persuasive *logoi*.) The known text versions of the story of the imprisoning throne and the return of Hephaistos to Olympus are found in sources dating well after Homer. The return of Hephaistos to Olympus is also the subject of vase paintings, the preponderance of which date from the mid-sixth through the fifth centuries BCE.[228] The very popular story of the return of Hephaistos to Olympus in the company of Dionysos and satyrs will be discussed further below.

Hera is a powerful and puzzling goddess. Her power, Homer has her tell us, resides in her incestuous marriage with her brother, Zeus, and in being his senior, "the eldest daughter of Kronos." But Burkert points out that Zeus had a stock epithet, "the loud-thundering husband of Hera," and it can be said that some of his own authority resides in his connection with Hera.[229] Hera was worshipped in three forms: as the virgin Hera Pais (or Parthenos); as the Fulfilled, Hera Teleia; and the Separated or Widow, Hera Chera. Her connection to Zeus includes the sexual and even seductive connotation that exists within marriage, but her image also encompasses virginity as a prerequisite to marriage. The stories of her efforts to hide her indecorous conception of Hephaistos with Zeus out of wedlock serve to underscore this theme. As the Separated one, Chera, she spends a period separated from Zeus—and indeed during this time she renews her virginity, as celebrated in the yearly ritual bathing of her statue at Nauplion. Harrison believes that "Hera has been forcibly married, but she is never really wife," and that "Long before her connection with Zeus, the matriarchal goddess may well have reflected the three stages of a woman's life; Teleia, full-grown, does not necessarily imply patriarchal marriage."[230] Indeed, Hera gives birth within

---

[227] Gantz, 75.

[228] Frank Brommer, *Hephaistos: Der Schmiedegott in der Antiken Kunst* (Mainz am Rhein: Verlag Philipp von Zabern, 1978), 10.

[229] Walter Burkert, *Greek Religion*, trans. John Raffan (Cambridge: Harvard University Press, 1985), 132.

[230] Jane Ellen Harrison, *Prolegomena to the Study of Greek Religion* (Princeton: Princeton University Press, 1991), 316.

marriage but not necessarily through the participation of a male. It may be noted, in the context of the failure of sexual reproduction in a divine marriage, that a similar couple to Zeus and Hera exists in Indian myth, Shiva and Parvati. Parvati alone forms Ganesha from the scrapings of her own dead skin. He is elephant-headed because of an act of violence by his "father" Shiva, who strikes off his head when Ganesha bars his entry to Parvati's private chamber. The head is replaced by that of the first available creature, an elephant with a broken tusk, whose mark Ganesha bears thereafter. In a related mytheme, when he is distracted from coitus with Parvati, Shiva's spilled seed falls on the earth and generates the Pleiades, recalling the generation of Erichthonios from the spilled seed of Hephaistos.

Hera's casting Hephaistos off Olympus and his revenge by means of the golden throne followed by their mutual accommodation seem most emblematic of their relationship. That, and the quality of anger they share. Hera's is hot and vengeful and at her worst she responds with swift and vindictive punishments aimed toward the victim, not the perpetrator—usually Zeus—to assuage wounds to her pride. (Io is such a victim, transformed into a cow tormented by a gadfly as punishment for Zeus's having seduced her.) Hephaistos' anger is also vengeful but sublimated, tricky, calculated, and effective.

## The Other Mother: Thetis

The motif of the flight of Thetis ushers in two significant appearances of Hephaistos in the *Iliad*. Thetis' flight may be taken as a counterpoint to Hephaistos' flightlessness. In fact, cast twice from Olympus, he falls rather than flies, plunging on the first occasion when Hera discards him into the ocean where he is rescued and raised by Thetis and Eurynome. "Silver-Footed" Thetis' beauty is also a counterpoint to "Crook-Footed" Hephaistos' ugliness. Yet it will be seen that they are linked by significant thematic affiliations.

Hesiod names Thetis as one of the fifty Nereids, daughters of the sea-god Nereus, son of Okeanus, who is one of the children of Gaia and Ouranos from the first generation of gods and the source of all waters. Thetis lives in the sea, either with her father or with Eurynome (one of the 3,000 daughters of Okeanus).[231] Hephaistos

---

[231] Gantz, 16.

spends nine years in Thetis' sea cave, tutored by Kedalion (the "phallic one") in metalsmithing and making exquisite objects (*daedala*) to delight his foster mothers.[232]

Although clearly not a god of the aether, Hephaistos nevertheless has powerful elemental affinities in addition to fire, and his connection with Thetis and her sea-cave links him with water. Indeed, the god is representing his own foster-lineage as well as Achilles' mother-line when he depicts Okeanus encircling the famous Shield. The *technê* of the god of fire and forge can be said to have been born in the mysterious Ocean. That Hephaistos can be described as joining the two elements of fire and water is another indication of his solidarity with the mythic blacksmith lineage and his connections to powerful forces whose sources lie outside Zeus's order.

Thetis and Hephaistos have another bond, that of mortal pain. Thetis nurtures infant Hephaistos when he is overcome by the "mortal pain" that follows his fall when Hera casts him from Olympus. While Hephaistos is separated from the other gods by his irreversible lameness, Thetis, who bursts into tears in Hephaistos' workshop, laments her own special distinction:

> "Oh Hephaistos--who of all the goddesses on Olympus,
> who has borne such withering sorrows in her heart?
> Such pain as Zeus has given me above all others!
> *Me* out of all the daughters of the sea he chose
> to yoke to a mortal man, Peleus, son of Aeacus,
> and I endured his bed, a mortal's bed, resisting
> with all my will. And now he lies in the halls,
> broken with grisly age...."[233]

Her present sorrow is, of course, on account of her half-mortal son Achilles, for the divinely wrought armor Hephaistos makes for him will ensure his fame but will not avert his fate, which is to be mortal and short-lived.

In Homer's world, the implacable will of Zeus has made brutal use of both Hephaistos and Thetis, yet both retain special gifts of persuasion—the power of *logos* (in its pre-Platonic sense, described in Chapter 2) that enables those who are weaker to prevail. Thetis has the power to grasp Zeus by the knees and persuade him to take action

---

[232] Homer, *Iliad*, trans. Fagles, 18: 466-73.
[233] Ibid, 18:501-08.

in Achilles' behalf, indeed temporarily against his own ultimate purposes. Limping Hephaistos' words have the power to calm Hera's anger at Zeus for granting Thetis this favor and to avert the wrath of Zeus from Hera and the Olympians assembled at the banquet table.

Like the Okeanid Metis and her own father Nereus (whom Herakles encounters and from whom he wrests information only after retaining his grasp through myriad shifts of form), Thetis is a shape-shifter.[234] Seventh- and sixth-century vase paintings show Thetis transforming herself into fish, snake, lion, panther and other forms to escape the grasp of Peleus.[235] (Examples include a 500 BCE *kylix* from Vulci[236]). Hephaistos is not a shape-shifter, but as has been shown, he is a possessor of *mêtis* and as an artisan he effects magical transmutation of form through his creations.

Thetis is a "minor" goddess, but Laura M. Slatkin points out the power of Thetis as "a figure of cosmic capacity, whose existence promises profound consequences for the gods."[237] Thetis is able to extract Zeus's promise to reverse the direction of war by virtue of her unique position as Zeus's rescuer when Hera, Poseidon, and Athena sought to overthrow and bind him. Binding is the only effective method of neutralizing the power of an immortal (as, for example, the immortal Titans are bound in Tartarus following their defeat in the battle between the gods and Titans.) Thetis summons Briareus, the "Hundred-Hander" to defend Zeus, "and the blessed gods feared him, and ceased binding Zeus."[238] Rescuer of Hephaistos and Zeus, Thetis also "receives and comforts" Dionysos when he is attacked by Lykourgos on Mount Nysa and dives into the sea to escape.[239] The source of Thetis' power thus to intervene is not directly stated. But, both Hephaestus and Thetis are among the gods that bind and unbind, as will be discussed below.

---

[234] Gantz, 16.

[235] Ibid, 229.

[236] Berlin F 2279, Perseus Vase Catalog, Perseus Digital Library Project, ed. Gregory R. Crane, 2004, Tufts University, accessed November 5, 2004, http://www.perseus.tufts.edu/.

[237] Laura M. Slatkin, *The Power of Thetis: Allusion and Interpretation in the Iliad* (Berkeley: University of California Press, 1991), 72.

[238] Ibid, 60-61.

[239] Homer, *Iliad*, trans. Fagles, 6.130-40.

That cosmic power resides in Thetis cannot be doubted. In a doubling of the mytheme of the prophecy that Metis will bear a son who will defeat his father, which is why Zeus swallows her, the same is also foretold of Thetis. Gantz recounts that the knowledge and ultimate divulging of this secret is what causes Prometheus to be reconciled to Zeus. It also causes Zeus to cease his amorous pursuit of Thetis and to dispose her in marriage to a mortal, Peleus.[240] Thus, Thetis' son, Achilles, if born to Thetis and Zeus would have threatened the rule of Zeus. The *Iliad* represents the close of the age of heroes recorded by Hesiod. Unlike their immortal parents, the half-divine sons of goddesses and gods—like Achilles—will die in glory and their fame will endure while their potential to affect the divine order will be firmly extinguished.

"The Homeric figure of Thetis retains older traditions of a divine creator-goddess...."[241] In Greek myth she is sometimes confused with her grandmother Tethys, the wife of Okeanus, the origin of *all things*.[242] So, too, Hephaistos is clearly representative of a more ancient mythos connected with creativity; and this is at least in part in through collaboration with his foster-mother, Thetis.

## Aphrodite and Charis

The fact that Hephaistos is married is in itself unusual among the Olympian gods—the only other god depicted as such in the Homeric epics is Zeus. Unlike Zeus or the other married Olympian god, Dionysos, Homer has Hephaistos married to two different goddesses: in the *Odyssey*, he is married to Aphrodite; in the *Iliad* he is married to Charis.

Gantz notes that on the "François Krater"[243] (ca. 570 BCE – 560 BCE), Aphrodite is depicted standing in front of Zeus at the arrival of Hephaistos on Olympus, which might suggest an interpretation that, "Possibly her special interest in the event stems from the assumption

---

[240] Gantz, 228.

[241] Sarah P. Morris, *Daidalos and the Origins of Greek Art* (Princeton: Princeton University Press, 1992), 81.

[242] Detienne and Vernant, 142, italics mine.

[243] Florence 4209, Perseus Vase Catalog, Perseus Digital Library Project, ed. Gregory R. Crane, 2004, Tufts University, accessed November 5, 2004, http://www.perseus.tufts.edu/.

that Hephaistos is here, as in *Odyssey* (Book 8), her husband, in which case her look probably denotes disappointment at his return." At least one scholar has suggested that the poet Alkaios hinted that Hephaistos demanded the hand of Aphrodite as the price for freeing Hera. Gantz, however, states that there is no evidence for this.[244] In other versions, Hera has effected the arrangement (perhaps both out of relief at being freed from the throne and pure spite toward Aphrodite who is by nature an enemy of the patriarchal valuing of marriage that Hera has come, by the time of Homer, to represent). That Hephaistos later demands back the bride-price he has paid to Zeus for Aphrodite gives some credence to the latter. However: how, when, why, and for how long Hephaistos and Aphrodite were paired in marriage remains a matter for speculation. The marriage is nowhere celebrated (as is, for example, the marriage of Peleus and Thetis); rather, the fact of the marriage is known mainly through the story of the adultery of Aphrodite and Ares as told in the *Odyssey*. In images, Aphrodite is not paired with Hephaistos as often as instead with Ares, and the two have children: Harmonia, Phobos ("Fear"), Deimos ("Terror").[245] In later accounts Eros, originally named by Hesiod as a primal god, "the limb-loosener who conquers the hearts of mortals and gods," is accounted as the child of Ares and Aphrodite.[246]

What then is the significance of Hephaistos' marriage with Aphrodite? Is it the union of *technê* with beauty in the aesthetic sensibility that endows objects, and by association their users, with an erotic glow—like the girdle Aphrodite lends to Hera that the wife may seduce the husband, Zeus, in order to have her way, in the bedchamber that "locked with a secret bolt no other god could draw"?[247] Is it instead the contrast between the boundaries of craft (remembering that *technê* came to mean the objective standards whereby a product of craft could be judged) and the unbounded, chaotic fecundity of nature? I would suggest the former. The ancient texts allow for either interpretation.

---

[244] Gantz, 76 ; 76 n. 22.

[245] Ibid, 80.

[246] Ibid, 3.

[247] Homer, *Iliad*, trans. Fagles, 14:204 ff.

On the two occasions in the Homeric epics that Hephaistos draws laughter from the Olympians, his lameness is specially mentioned. In the case of the gods' banquet, the laughter is drawn by Hephaistos' soothing *logoi* and the apparent self-mockery of his shuffling gait as he usurps the cupbearer's role and mimics graceful Ganymede and gets away with it.[248] In the case of the golden net, the Olympians laughingly remark that the swiftest of gods, Ares, has been snared by the slowest.[249] Does the lameness of Hephaistos separate him from the delights of physical love? The texts do not state whether this is so or not. This question will be further considered.

What I believe to be the more significant question in the Homeric context is why does Hephaistos *divorce* Aphrodite? For this is what he does by demanding her bride-price back from Zeus. Hephaistos calls her "dog-eyed" (*kunôpês*, meaning "shameless")—the same term that he uses for Hera in the *Iliad*, often translated as "bitch."[250] This negative animal appellation seems mild compared with certain "fierce" Aphroditic appellations noted by Roscher and Kerényi:

> the dark one," or "the black one," associates her with the three-faced figure of Hekate of whom the witches are fond and to whom dogs were sacrificed, and also the terrible Erinyes among whom she was named as one. The goddess of delicacy and roses was also called *androphonos* (killer of men) and *anosia* (the unholy) and *tymborychos* (the gravedigger). There is also a black-bearded Aphrodite....[251]

James Hillman also notes that in Sparta and Corinth, "there was a local cult of warrior Aphrodite."[252] If (with the exception of his sole appearance on the battlefield of Troy when he is called out by Hera to fire on the River Xanthus) Hephaistos is known as a peacemaker, the match between Hephaistos and Aphrodite does not seem to make sense aside from its comic value in the *Odyssey* (where it serves perhaps as a counterpoint for the tale's hearers/readers, who know that Odysseus has a faithful wife back at home). Although Homer makes Hephaistos complain bitterly about the misfortune of birth that

---

[248] Burkert, 168.
[249] Homer, *Odyssey*, trans. Fagles, 8: 373-74.
[250] Homer, *Iliad,* 18:396, *Homeri Opera.*
[251] James Hillman, *A Terrible Love of War* (New York: Penguin, 2004), 144.
[252] Ibid, *Love of War*, 144.

made his legs crooked in contrast to Ares' straight, racer's legs, it must also be remembered that once Hephaistos secures Zeus's agreement to the return of the bride-price and Poseidon's guarantee of Ares' payment of the adulterer's reparation, he releases the pair without further complaint or comment, as if the cash transaction is what matters.[253] I doubt it. Excellence and magic in making is what matters. As for Aphrodite, following her release from the golden net, only the sound of her silvery laughter is heard to reveal the direction she has taken, back to her cult-center in Cyprus—where, it must be mentioned, she will be attended as always by the Graces. In taking one of the Graces for wife, does not Hephaistos still claim possession of the highest degree of aphroditic charm?

To the degree that the poetic use Homer makes of Hephaistos is to represent and to some measure affect the order of the Zeusian universe, his union with disorderly Aphrodite is unnatural. That the pair generate no children would follow from this. The question this scene should also raise is not so much whether Hephaistos and Aphrodite are properly paired, but rather that Aphrodite and Ares are.[254]

What then of the marriage of Hephaistos and Charis in the *Iliad*? Charis is not a proper name, but rather means, simply, "Grace." Hesiod indicates that Aglaia is the one of the three Graces who is married to Hephaistos.[255] Homer paints this marriage as harmonious and Grace as a seemly and beautiful helpmate to her busy husband, glimpsed as she welcomes Thetis to their home, which is also the workshop of Hephaistos. Here, technology is harmoniously married to beauty, gentleness, and grace—both personal and social, as represented by Charis—and not to energetic lust, as represented by Aphrodite. Nevertheless, marriage to either or both of the wives Homer gives to Hephaistos paints the outward portrait of grace married to ugliness. There is always a deeper story in such unions. In Chapter 4, I will return to the central question of the relationship of phallic Hephaistos with the chthonic feminine.

---

[253] Homer, *Odyssey*, trans. Fagles, 8: 397-401.
[254] See: James Hillman's *A Terrible Love of War* on the profound connection between Ares and Aphrodite expressed in the emotions of war.
[255] Gantz, 76.

## Gaia and Athena

Hephaistos, smitten with desire for his magnificent sister, advances on Athena (who, in at least one version, has come to his workshop to select a spear). She resists and he intemperately ejaculates on her thigh. She brushes off Hephaistos' semen with a piece of wool and casts it on the ground (with a moue of distaste—or perhaps with the knowledge and intent that the divine seed will prove fertilizing), whereupon Gaia bears a child. Numerous vase paintings, for example a *stamnos* dated c. 460 BCE, show Gaia rising from the ground to deliver this child, Erichthonios, into Athena's arms.[256]

Nicole Loraux remarks that, "As the 'midwife' for Zeus at the birth of Athena, the maker of Pandora, and the father of Erichthonios, Hephaistos seems predisposed to intervene in all generation that occurs without sexual reproduction."[257] This assertion of Hephaistos' peculiar relationship in regard to sexual reproduction is justified. In the next chapter I will show that this is an important feature of the chthonic phallic energies closely associated with the god (especially readily seen in the image of the folkloric dwarf). Loraux further notes that Marie Delcourt "aptly observes that he occupies this place by virtue of being a son of a mother who has no husband."[258] This requires a bit more unpacking. Below, I will discuss the work of Delcourt and others on the relationship of Hephaistos to the oldest generations of gods, for whom fecundity is not necessarily reliant on sexual reproduction, and whose powers, including their powers of production and reproduction, are often linked with the Earth goddess, Gaia. Like Hera, they may bring forth monsters like Typhaon. It should also be observed that monsters or combined forms, like Erichthonios, sometimes described as snake-tailed, and born of the spilled seed of Hephaistos, are emblems of primordial powers.

For the Athenians, Erichthonios, whose name reflects his birth out of the earth as first citizen and ancestor, symbolized their rootedness in the very soil of the city and their special status as citizens. Indeed,

---

[256] Munich 2413, Perseus Vase Catalog, Perseus Digital Library Project, ed. Gregory R. Crane, 2004, Tufts University, accessed November 5, 2004, http://www.perseus.tufts.edu/.
[257] Nicole Loraux, *The Children of Athena: Athenian Ideas About Citizenship and the Division Between the Sexes* (Princeton: Princeton University Press, 1993), 128.
[258] Ibid, 128.

"From the perspective of an official narcissism, there is, in effect, no citizen who is not an autochthon (*auto-chthôn*: born from the soil itself of the fatherland)."[259] In fact,

> On the Acropolis, the Athenians celebrate two remarkable births, and the myth places them both under the joint patronage of Athena and Hephaistos—the birth of Erichthonios, and the birth of Pandora: the first of the Athenians [the] authochthonous figure who made the goddess, his protector, into the eponym for the city, and the first woman, an artifice.[260]

The birth of Pandora was represented on the base of the statue of Athena in the Parthenon. Loraux offers a complex and compelling thesis on Pandora, the "artifice," as the genetrix of the "race of women" (based on Hesiod's version of the myth of woman's separate creation as a bane to men) and the expression of this idea in the status of Athenian women, who, considered to be engendered in a separate creation, were in effect denied the human status accorded to men. When initiated, their sons were no longer the sons of mothers but regarded as descended from the Athenian earth. Loraux's thesis relies in part on the necessity for Pandora to be viewed as an *artificial* creation, as distinct from the "birth" of Erichthonios from the seed of Hephaistos which fertilizes the generative Earth.[261] As will be discussed in Chapter 5, Murray Stein similarly asserts that the "unnatural" creations of Hephaistos are a psychic affront to naturally generative female creativity. I believe that both views are overstated in regard to the negative evaluation placed on the artificiality of Hephaistos' creations relative to the 'naturalness' of other modes of generation. I intend to argue that the creativity of Hephaistos has very different meanings, rooted in primordial images that can be said to precede as well as exist alongside images of sexual reproduction.

In this connection it must also be noted that the variants of Hephaistos' marriage and family relations include a memory of Hephaistos married to Athena. In another version of the aftermath of Hera's release from the golden throne, Zeus offers Hephaistos a gift in return for freeing Hera "and on Poseidon's spiteful advice"

---

[259] Ibid, 37.
[260] Ibid, 88.
[261] Ibid, 88 n. 99.

(Poseidon and Athena having battled over the possession of Attica), Hephaistos asks not for the hand of Aphrodite but of Athena.[262]

Whether or not married, Athena and Hephaistos are in partnership, especially in Athens, where they were jointly revered as the patrons of artisans. It is Hephaistos *together with Athena* who is honored as the giver of all good things in the "Homeric Hymn to Hephaistos" (c. 450 BCE). Their statues likely stood together in the temple to Hephaistos, the Hephaisteion, in Athens, and Hephaistos was accorded an altar in the Parthenon. It is in Athens that Hephaistos' most overt cult recognition occurred (his connections with mystery cults on Lemnos and elsewhere will be discussed in the next chapter). Loraux observes that Hephaistos is without question the lesser partner to Athena, just as Hera laments that,

> Without me, [Zeus] has given birth
> to bright-eyed Athena,
> who stands out from all the blessed gods.
> But my own boy, Hephaistos,
> the one I myself gave birth to,
> was weak among all the gods....[263]

It is clear that Athena is regarded as the favored daughter of Zeus and the virile virgin who represents the glorious Athenian *polis* to itself. And, "the birth of Hephaistos must take second place to that of Athena."[264] The birth of the goddess without a mother supersedes that of the god without a father. Loraux points out that although the birth of Hephaistos is never depicted, such is of course not the case with the birth of Athena. At this event, however, Hephaistos is virtually always depicted, for example in an Attic Black Figure tripod *kothon* from Thebes.[265] Loraux asserts that "Hephaistos' role in the affair is clear:" in cleaving the head of Zeus with his hammer, "he is crucial as an instrument of liberating magic."[266] This then was a key to his cult position in Athens.

---

[262] Gantz, 75.

[263] "The Hymn to Pythian Apollo," 168-169.

[264] Loraux, 131.

[265] Louvre CA 616, Perseus Vase Catalog, Perseus Digital Library Project, ed. Gregory R. Crane, 2004, Tufts University, accessed November 5, 2004, http://www.perseus.tufts.edu/.

[266] Loraux, 131.

The Hephaisteion, erected in Athens in the mid-fifth century BCE, presented its face to the artisans' quarter.[267] Events honoring Hephaistos were sponsored by the artisanal cooperative of Athens. The artisans of Athens were the best organized of any Greek city.[268] Hephaistos is related to Athena through artistry, craft, and work (remembering that one of Athena's titles, Ergane, signifies "Athena the Worker"). They are also related through the quality of *mêtis*. As Atchity remarks, the "'ingenious' (*polúphronos*) nature of the golden snare [Hephaistos] fashions for Ares and Aphrodite is equivalent to the 'crafty' (*dólos*) character of the Trojan horse, built by Athena-inspired Odysseus."[269] Though both possess *mêtis*, without doubt their relationship to work is of a different and complementary character, Athena's more strategic, educational, and inspirational, Hephaistos' more grounded and instrumental. Their key collaborative work is Pandora.

## Pandora

It is the story of Pandora that presents Hephaistos in the role of creator at the behest of Zeus, molding earth to form the first woman. Pandora's story is told twice by Hesiod. In the *Theogony*, the creation of Pandora is an evil plotted for men, on whose behalf Prometheus has stolen fire. Hephaistos molds earth to form her and Athena dresses the maiden and places on her head a wondrous crown created by Hephaistos. An assembly of gods and men wonder at her beauty. In *Works and Days* the story is told again, with more detail. Prometheus has deceived Zeus, who withholds fire, which Prometheus steals from him and hides in a fennel stalk as a gift to men. In revenge, Zeus orders Hephaistos to mix water and earth to form the maiden, and to give her the face of the goddesses together with voice and strength. Other gods then give her various gifts: Athena teaches her skills such as weaving, Aphrodite gives her *charis*, "grace," and the power to incite desire, Hermes gives her a thievish and deceptive character. Hermes is sent to deliver Pandora to

---

[267] Vincent Scully, *The Earth, the Temple, and the Gods*, Rev. ed. (New Haven: Yale University Press, 1979), 189-90.

[268] Delcourt, 26.

[269] Atchity, 135.

Epimetheus ("after-thought"), brother of Prometheus ("forethought"). Through her union with the Titan Epimetheus, she will be the origin of the "race" of women—who could be said to be a different race from men, through a separate and presumably later genesis than that of men. Hesiod's telling of the story is notoriously misogynistic.

Gantz points out that extremely little of this material appears in the two centuries following Hesiod and reports "a variant in which Prometheus gets the jar containing Pandora's gifts from the Satyroi," suggesting reference to a lost satyr play, perhaps Aeschylus' *Prometheus Pyrkaeus* ("Firekindler"), that "burlesqued Prometheus' gift of fire to the first men (i.e., Satyroi)" or Sophocles' *Pandora*, alternately known as *Sphyrokopoi* ("Hammerers").[270]

Jane Ellen Harrison identifies Pandora as the Kore form or title of the Earth-goddess.[271] The Kore, or Maiden, also represented by Persephone, belongs to the underworld.[272] Pandora's story also contains the image of the jar (*pithos*) or great vessel she presumably brings with her, though its origins are not given. She removes the lid, releasing evils, illness and human cares into the world, only stopping the jar in time to trap Elpis (usually translated as "Hope") within.[273] Harrison sees in the jar a remnant of the grave-*pithos* of the chthonic goddess.[274] Pandora's earth-connection is evidenced in a series of vase paintings showing a youthful female figure rising from the ground. This iconology of the female figure rising halfway out of the ground is usually associated with Gaia (as in the images of her rising out of the ground and presenting Athena with Erichthonios) but in at least one example a youthful figure is labeled "Pherophatta," or the "Anodos of Kore." This rising figure is surrounded by Panes (goat-men), and in another example by horse-tailed Satyroi.[275] In some representations featuring the rising Kore, Satyroi or Panes carry picks or hammers of the kind employed by Hephaistos to break open the head of Zeus. In at least one example, an Attic Black Figure *lekythos* reproduced in Harrison, men with vines radiating from their heads

---

[270] Gantz, 157, 158, 162.
[271] Harrison, 281.
[272] Ibid, 276.
[273] Gantz, 157.
[274] Harrison, 282, 285.
[275] Ibid, 277-278.

raise their hammers as if to strike or furrow the large head of a woman rising out of the earth.[276] In one vase painting specifically identifying the rising figure as Pandora, Epimetheus, her prospective husband, who presumably has just seen her for the first time and over whom hovers an Erote, carries a hammer.[277] The content of Aeschylus' "Hammerers," dealing with Pandora, is unknown. The hammer characteristic of Hephaistos, according to Harrison, was also used by agriculturalists in breaking earth (the spade being a later innovation). Harrison recalls a ritual performed at the Delphian festival of the Charila, held every nine years, in which "a puppet dressed as a girl was brought out, beaten, and ultimately hanged in a chasm," a ritual sacrifice to purify the community against the effects of a long-ago famine. Could the Satyrs with their picks and hammers be a representation of "storm and lightning beating on the earth to subdue it and compel its fertility"? Are they raising her from the earth? Or in imitation of Hephaistos, are the hammerers making Pandora?

In any event, the identification of Pandora with the Earth is supported by bits of testimony identifying Pandora as an earth-goddess to whom older rituals were addressed that by the time of Aristophanes was remembered only formulaically. In a "late" vase-painting depicting her birth,

> She no longer rises halfway from the ground, but stands stiff and erect in the midst of the Olympians.... Earth is all but forgotten, and yet so haunting is tradition that, in a lower row, beneath the Olympians, a chorus of men, disguised as goat-horned Panes, still dance their welcome. It is a singular reminiscence, and, save as a survival, wholly irrelevant.

In Hesiod's misogyny, Harrison detects "the ugly malice of theological animus" in the universe of Zeus, who will not permit the hegemony of the great Goddess:

> Zeus the Father will have no great Earth-goddess, Mother and Maid in one, in his man-fashioned Olympus, but her figure is from the beginning, so he re-makes it; woman, who was the inspirer, becomes the temptress; she who made all things, gods and mortals alike, is

---

[276] Ibid, 279, fig. 69.
[277] Ibid, 281, fig. 71.

become their plaything, their slave, dowered only with physical beauty, and with a slave's tricks and blandishments.[278]

One could here for "blandishments" substitute "*logoi*." Like the Panes in the (literal) lower row, the story of Pandora points out that in the character of Hephaistos something too remains below the surface of Zeusian order that recalls earlier associations and origins linked with the Earth.

The "birth" of Pandora was represented on the base of the statue of Athena in the Parthenon. Loraux observes that,

> The Hesiodic myth had nothing to say about the origin of men, while the creation of woman, that artifice, was a purely artisanal affair, derived only from a single discourse.... By adopting Pandora, the city of Athena accommodated the race of women, but denied the existence of a "first Athenian woman.[279]

As will be discussed below, Hephaistos no less than Pandora or indeed Athena, was co-opted into the service of the Athenian *polis'* idea of itself.

Prometheus and Zeus: The *Technê* of the Maker and the *Technê* of the *Polis*.

Both Hephaistos and Prometheus were revered by the Athenian *polis* for the creation of good things for human life. The "Homeric Hymn to Hephaistos" praises him as the god whose skills, taught to humankind, allow mortals to live in comfort the year round in well-built homes. In Aeschylus' *Prometheus Bound*, the gifts that ease and elevate the condition of humankind are attributed to Prometheus— "architecture, advanced methods of agriculture, the domestication of animals, divination and writing."[280] It is unknown whether Aeschylus originated these services to humanity on the part of Prometheus; they are not mentioned by Hesiod in whose *Theogony* and *Works and Days* Prometheus makes his first appearance.[281] He does not reappear in extant texts until the fifth century in Athens.[282] In Athens, both

---

[278] Harrison, 282-85.
[279] Loraux, 10.
[280] Gantz, 159.
[281] Ibid, 159.
[282] Morris, 360.

Hephaistos and Prometheus were honored in the festival known as the Prometheia.

The similarities between Hephaistos and Prometheus have to do with their relationship to technology. Their myths overlap. However, it is the difference between them, including the differences in their respective relations with Zeus, that will help, through contrast, to sharpen the picture of Hephaistos discernable in Greek myth.

Hesiod's Prometheus is one of the Titans (the second generation, the first having been defeated and bound in the eternal darkness of Tartarus). He runs afoul of Zeus at the occasion of the first sacrifice. A great cow is to be divided by men and gods. Zeus chooses the bones as the gods' share, which Prometheus has disguised by wrapping them in rich and fragrant fat. The meatless bones are henceforward the gods' share of the sacrifice. Angered by this deception, Zeus's revenge is to withhold the gift of fire from humankind, which Prometheus then steals, hiding it in a fennel stock. This is when Zeus conceives the idea of the maiden Pandora, the first woman, to be made by Hephaistos and then sent to bedevil men by mothering the race of women, who will drain men of their prosperity through laziness and malicious temper.[283] Zeus chains Prometheus to a column with wondrous bonds—these bonds are made and installed by Hephaistos—and sends an eagle to tear at his immortal liver, which restores itself each night for the next day's tribulation. Eventually, Herakles slays the eagle (with Zeus's permission), but, like the unremitting punishment of the Titans, Prometheus remains bound. Gantz comments that, for Hesiod, "so long as Prometheus, like Atlas and the Titans in Tartaros, remains under restraint, the reign of Zeus, and the order which it brings, will be better secured."[284]

According to Aeschylus (and fragmentary accounts of his Promethean trilogy, of which only *Prometheus Bound* survives) Prometheus possesses the secret of the name of the goddess— Thetis—fated to bear a son greater than Zeus. It is for this reason that Zeus threatens Prometheus with the torment of the eagle if he does not reveal it. Prometheus is released when he reveals the secret and prevents Zeus from consummating his pursuit of Thetis, and

---

[283] Gantz, 155.
[284] Harrison, 156.

Prometheus and Zeus are reconciled. Even after his release, Zeus "caused the newly freed Prometheus to wear a garland as a substitute for his imprisonment."[285] Elsewhere, the object is a ring (which would naturally suggest Hephaistos as its maker). Clearly, Prometheus remains bound by and to Zeus, in both enmity and amity.

Prometheus is also more clearly on the side of humanity, credited with saving men when Zeus threatens to destroy them. Such is not so clearly the case with Hephaistos, whose status is that of an Olympian god. Hephaistos, too, has been punished by Zeus. Both have trickster traits (Prometheus at the sacrifice and as thief of fire, Hephaistos as maker of Hera's imprisoning throne and the golden net), but unlike Prometheus, Hephaistos is not subjected to binding by Zeus. Hephaistos supports the order of Zeus through his creations, but also voices dissent. When Zeus's henchmen Kratos (Power) and Bia (Force) accompany Hephaistos, truculently urging him to his task of binding Prometheus to the rock, although the *temenos* of Hephaistos has been violated—Prometheus has stolen fire for humankind from the temple of Hephaistos and Athena—Hephaistos is vocal in his reluctance to carry out Zeus's order. Aeschylus places in Hephaistos' mouth a lament at Zeus's sternness toward Prometheus. As a "new tyrant," Zeus needs to consolidate power by all means necessary. Hephaistos' moral ambiguity in this scene—he registers his protest but carries out the work—echoes his response to Hera's call to fire on the River Xanthus, which he performs impersonally, as another job of work.

In Plato's retelling of the Prometheus myth (discussed in Chapter 2), a sharp distinction is drawn between Prometheus' gifts to humanity of *demiourgikê technê*, the arts of making (including the teachable *technai* of Hephaistos and Athena that make the conditions of mortal life bearable and good) and Zeus's gift of *technê politikê*, the necessary political arts which regulate the conditions of human interaction. The making arts alone will not sustain the good qualities of human life; instead, their instrumental power being greatly enhanced by Prometheus' gifts, men "injured one another for want of political skill."[286] Only when regulated by the gifts of Zeus can the

---

[285] Ibid, 160.
[286] Plato, *Protagoras*, 322b, *Collected Dialogues*.

gifts of Prometheus cause humans to flourish and distinguish themselves from animal existence: "'To all,' said Zeus. 'Let all have their share. There could never be cities if only a few shared in these virtues, as in the arts'" (and Zeus further stipulates that, "if anyone is incapable of acquiring his share of these two virtues he shall be put to death as a plague to the city").[287] Thus, Plato makes the *demiourgikê technê* of Prometheus subordinate to and dependent on the organizing principle of *technê politikê*, the province of Zeus alone. Prometheus' gifts are thus bound inextricably within the encircling order of Zeus. The *technê* of Hephaistos, however, is not so clearly categorized.

### Hephaistos, Daidalos, and Divine Artifice

Hephaistos is at times confused or inter-identified with Prometheus as the bringer of good *technai* to humanity. Such is also the case with Daidalos, the legendary maker of the Cretan labyrinth and the wings of Ikaros. The most important place-connection of Daidalos is of course to Crete. Greek myth gives Hephaistos some Cretan connections, including the creation of a bronze man, Talos, to serve the Cretan king, Minos.

The first certain appearance in literature or art of Daidalos occurs in Homer, in Book 18 of the *Iliad*, in the description of the making of the Shield of Achilles which culminates in the spectacle of leaping dancers:[288]

> And on it the renowned, ambidextrous artist inlaid a dance,
> Like the one which once in broad Knossos
> Daidalos crafted for Ariadne of the lovely hair.[289]

As will be shown, "*daidalic*" making, specifically associated with Hephaistos in epic poetry, contains both magical and fateful qualities. Recognizable through their profoundly artful elaborateness and lifelikeness, the appearance of *daidalic* objects signals themes of fate and divine intervention. Sarah J. Morris observes that though Daidalos is mentioned only once by name, the "daidalic" is alluded to throughout the *Iliad* by Homer's use of cognates of words deriving from a root, *-dal*, whose etymology and exact meaning is

---

[287] Plato, *Protagoras*, 322d, *Collected Dialogues.*
[288] Homer, *Iliad*, trans. Fagles, 18:592.
[289] Morris, 13.

undetermined although Indo-European and Semitic roots have been proposed. Reduplicated as *daidal-* to produce adjectives, a noun, an infrequently used verb, and the name of Daidalos himself these words have a very special purpose in Greek epic poetry. They

> describe, represent, or personify objects of intricate and expensive craftsmanship; expressions such as "well-crafted," "intricately worked," or "skillfully wrought" will satisfy their meaning, encouraged by two instances as a verb for the activity of a craftsman at work.

In fact, the name Daidalos is one of several eponymous names impersonating a class of activity used in Homer, such as "the builder," "the joiner," "all names for artists derived from their activity." Even the name "Homer" itself may exemplify this kind of naming.[290]

The most frequent use of the *daidal-* set of words describes arms and armor, though never of an ordinary kind, but consistently associated with the concept of the *kleos*, "glory," of heroes and deeds of epic heroism, as well as with the honoring and burial of heroes. Indeed, the epithet *daidaleos* used to describe objects associated with them contributes to the heroic image of the characters such as Achilles, and is seen as much in scenes of the hero's arming as in battle. Nevertheless, these objects retain a sense of identity independent of the owner; if anything, retaining a reflection of their maker. Indeed, the glamour of these objects outlasts the life and action of their owners, retaining a unique value and luster of their own. There are two examples of objects described as *daidaleos* that are associated with Achilles. One is the armor and Shield made by Hephaistos. The second is the chest, *kalos daidaleos* ("beautiful and elaborate"), given to him by his mother Thetis, containing a ritual vessel, "exclusive to the most solemn of ceremonies for the greatest of the gods," which Achilles uses to entreat Zeus on behalf of his friend Patroclus. No more will the *daedala* made by Hephaistos save Achilles from his fate than will the contents of this chest save Patroclus. The *kleos* Achilles entreats from Zeus for his friend will be awarded "in the form of death in battle." As Morris points out, "what attracted the adjective were its associations rather than its appearance,

---

[290] Ibid, 3-4.

beginning with its divine donor. For Thetis gave it to her son Achilles, to much the same effect brought on by her gift of a new suit of armor: the death of a hero."[291] Thus, daidalic objects are "glamorous but treacherous, qualities borne out in all their appearances" in epic poetry.[292]

Armor is not the only *daidalic* association in poetry. Many objects are mentioned in the *Iliad*, such as the *daedala*, presumably gold ornaments, with which Hera adorns Aphrodite's girdle, made by Athena, as part of her preparation to seduce Zeus in Book 14 of the *Iliad*.[293] Although describing ornaments and not arms, Morris remarks that Hera's "strategic toilette" is:

> in essence an arming scene in drag: a goddess prepares to seduce, hence conquer, the king of the gods in order to save the Greeks. Hera's role is no less vital than that of Patroklos or another hero whose entry into battle turns the tide of destruction, and her preparation is no less painstakingly described.[294]

Morris notes that Pandora, too, is laden with "a panoply of epic embellishments," the most striking of which is the golden crown made for her by Hephaistos:

> And he worked on it many *daedala*, a wonder to see,
> Wild beasts, all that the earth nourishes and the sea,
> of these he put many on it, and much grace shone from it,
> wonderful creatures, like living creatures *with voices*.[295]

The use of this "word complex" is at once idiosyncratic and intensive; and Hesiod reserves the use of these emphatic words exclusively for describing the attributes of Pandora, the ultimate "beautiful evil."[296] So lifelike are the creatures on Pandora's crown that they seem to have voices, recalling Homer's *ekphrasis* on the Shield which conveys the notion that a shepherd depicted on it is even audible in his singing.

*Daidaleos* is sometimes applied to portentous objects made by named mortal craftsmen. A rare instance of the adjective being

---

[291] Ibid, 18.
[292] Ibid, 17.
[293] Ibid, 19.
[294] Ibid, 20.
[295] Hesiod, *Theogony* 581-84, qtd. in Morris, 31.
[296] Morris, 31.

applied to foreign objects is a gift among the lavish prizes Achilles assembled for the funeral games of Patroclus, a "magnificent vessel,"

> a krater made of silver. It held six measures
> to pour, and in beauty it surpassed much for all time,
> since Sidonians of many skills had made it well (πολνδαίδαλοι),
> and Phoenician men had brought it across the wide sea,
> and set it up in harbors, and gave it as a gift to Thoas.[297]

Homer's use of the same adjectives to describe objects made by the god Hephaistos is significant. As in the case of another example, the foreign-made armor of Agamemnon, it indicates the Greeks' admiration for foreign objects of high aesthetic quality. "The convergence of these qualities—the *daidalic* and the exotic—is no coincidence, for it dominates Greek taste in art in the prehistoric as well as the epic and archaic world."[298] This admiration will undergo a political transformation in the re-imaging of Athens that will be seen to have a bearing on the transformation of Hephaistos' myth, together with that of Daidalos.

The appropriation of Hephaistos as ancestor of the Athenian *polis* in connection with the importance of *technê* was discussed above. The myth of Prometheus was revived in fifth-century Athens. Morris notes that,

> It was in fifth-century Athens that Daidalos became a sculptor, an Athenian, a relative of Hephaistos, a protégé of Theseus, and the hero of a local community. The cognitive forces that inspired this transformation shaped not only the legend of Daidalos but the Greek sense of history and self.[299]

Even Socrates claims Daidalos as an ancestor, in a genealogy linking *Socrates* with remote and famous ancestors backward to Hephaistos and Zeus.[300] Morris traces the course of the Athenians' rewriting of their own mythic history following their decisive and miraculous victory over the Persians at Marathon (490 BCE). What in the early aftermath invoked a pious attitude of gratitude for divine assistance, kept alive in civic oratory, gradually developed into a belief that "the Greeks, and Athenians in particular, were naturally

---

[297] Homer, *Iliad* 23:257-61, qtd. in Morris, 21-22.
[298] Morris, 22.
[299] Ibid, 26.
[300] Ibid, 257.

and culturally superior, hence able to overcome an enemy lacking the advantages of democracy and other Athenian institutions;" a belief that found representation in art and ambitious civic monuments. Hephaistos, Prometheus, and Daidalos were all appropriated as symbols of Athenian cultural and political authority, in a transformation whereby "mythology becomes a historical process in the classical era, transcending other 'universal' theories of myth (e.g., structuralism), which might otherwise apply to other periods of Greek culture."[301] Indeed, unlike Harrison, Morris discerns in the rising of Pandora from earth—seen in numerous examples from fifth-century Athens—not so much an unconscious representation of the ancient Mother as much as a conscious underlining of the autochthonous self-image of Athens.[302] The torch races known to have been associated with Hephaistos' and Prometheus' cults in Athens may have less to do with divine fire *per se* than with the legend of an Athenian runner who encountered the god Pan and brought his blessings and a message of solidarity back to the city on the eve of the Marathon victory. This encounter is said to have been the origin of the torch race, or relay, that was instituted for Pan in Athens at the Panathenia. The famous passage from Herodotus used by the U.S. Postal Service compares the Persians' postal runners to the runners in the torch race that was part of the cult celebration of Hephaistos in Athens: "Nothing stops these couriers from covering their allotted stage in the quickest possible time: neither snow, rain, heat, nor darkness."[303]

The most significant quality of the work of Daidalos how lifelike it is. In Athens, he was celebrated for making magic statues. In the earliest extant mention of such a work by Daidalos in a satyr play of Aeschylus, the satyrs exclaim,

> This image is full of my form
> This imitation of Daidalos lacks only a voice
> [....]
> It would challenge my own mother!
> For seeing it she would clearly turn and [wail]
> Thinking it to be me, whom she raised.
> So similar is it [to me].[304]

---

[301] Ibid.
[302] See: Morris, chapter 11.
[303] Morris, 320.

By the fifth century, philosophical speculation on the nature of humankind imagines human ancestors as a sort of work of art, to which the addition of a voice denotes the manifestation of spirit. Centuries earlier, Pandora's acquisition of "voice and strength" attributed by Hesiod to the art of Hephaistos made her a woman (though, Loraux argues, by Athenian standards not perhaps fully "human"). The girls in Hephaistos' workshop, though made of gold not clay, have wits, voice, and strength. Recall too Homer's ekphrastic evocation of the sound made by the boy piper forged on the Shield of Achilles. Daidalos' legendary works also gain such lifelikeness and indeed sentience that they appear to see as well as move. In plays, a statue of Pan mysteriously disappears, provoking the wonder, "Was it made by Daidalos, or did someone steal it?" Another speaks: "I am Hermes with a voice from Daidalos / made of wood (but) I came here by walking on my own."[305] Such is the expressive quality attributed to the work of Daidalos that it was said in Athens that his statues had to be chained down to prevent them from walking away.

However much fun Athenians had with the burlesquing of mobile satyric statues, and apparently they had a great deal of fun with it— the image nevertheless affirms the very real notions of divine images possessing such numinosity that they must be bound, either to retain their blessings or restrain their dangerous power. A famously chained statue of Artemis of Ephesos was credited with sending men mad "as a result of encountering her gaze."[306] Such numinous qualities were solely associated with objects made by a god like Hephaistos, a semi-divine culture hero like Daidalos, or the daemonic Telchines, said in some traditions to be the first makers of images of the gods. Ultimately, Aristotle imagines "instruments that would eliminate the need for slaves:"

> For if each instrument were able to accomplish its own task, either in obedience or anticipation [of others], like the [works of] Daidalos or the autokinetic tripods of Hephaistos which, as the poet says, "of their own accord entered the assembly of the gods," then in the same

---

[304] Ibid, 218.

[305] Ibid, 221-22.

[306] David Freedberg, *The Power of Images: Studies in the History and Theory of Response* (Chicago: University of Chicago Press, 1989), 75.

manner, shuttles would weave and plectra play the lyre on their own, master builders would not need apprentices, nor masters, slaves.[307]

## Gods Who Bind, Gods Who Are Bound, Gods Who Fall, Primordial Gods

The chaining of the statues of Daidalos recalls the ensnaring/immobilizing by Hephaistos of Ares and Aphrodite in the net made of magically invisible chains. Even ocean-born Aphrodite, respected for her sinuous *mêtis* (*aiolómêtis*) is taken by surprise. For Aphrodite sets snares and "traps" in her "insatiable desire to deceive and seduce, [making] Aphrodite a goddess to be feared among the gods as well as among men."[308] Hephaistos is, par excellence, among the gods who binds. In memory of the ensnaring net, a Greek proverb stated that "anything that cannot be avoided is called a bond of Hephaistos."[309] Aeschylus uses the word "net" to refer to the inextricable bonds Hephaistos placed on Prometheus.[310] The ring mentioned above in connection with Prometheus' symbolic perpetual binding to Zeus even after his release from the mountaintop would probably have been the kind the Greeks called "rings without limits" that "carry no stone or setting and so have no end or beginning."[311] Hephaistos likely made many of the same kind, along with the spiral coils Homer tells us were among the *daedala* he made in the undersea cavern of Thetis and Eurynome.[312]

Many gods and immortals suffer binding. The Titans are said to have been "bound" in Tartarus, a place under the sea of eternal darkness, without limits.[313] Zeus himself is bound by Poseidon, Hera, and Athena, a binding which Thetis forces them to undo by summoning from the bottom of the ocean Briareus, the Hundred-Hander, who bound the Titans in Tartarus. Thetis is "bound" in marriage to Peleus (a bond which we symbolize by wearing the circular ring without beginning or end). Hera and Ares are bound.

---

[307] Aristotle, *Politics* 1.4, qtd. in Morris, 225.

[308] Detienne and Vernant, 285.

[309] Ibid, 284.

[310] Ibid, 295.

[311] Detienne and Vernant, 292.

[312] Ibid, 300.

[313] Ibid, 293.

Hermes is threatened with binding. Dionysos the Loosener cannot be bound.

Another story of chaining involves Eurynome, who is credited, together with Thetis, with Hephaistos' rescue and fostering under the sea. Eurynome is one of the 3,000 Okeanids; but in another lineage she is even more primordial:

> Together with Ophioneus or Ophion, an Old Man of the Sea resembling Proteus, Nereus or Triton, she reigned over the world with her husband until Kronos and Rhea dethroned the ancient couple of the sea by making them *fall*, in the course of a struggle, *from the height of the sky to the depths of the ocean.*[314]

This ancient mytheme is echoed in the fall of Hephaistos. This Eurynome had a temple at Phigalia which was closed and secret. It was opened on only one day of the year, when an ancient *xóanon* (sculptural image or natural object depicting divinity) could be seen, depicting the goddess as half woman, half fish and chained in bonds of gold.[315]

The curious image of binding is associated with *mêtis*. There is far more to be said beyond the scope of this dissertation about the character of *mêtis* that expresses itself in bonds.[316] Detienne and Vernant also present a lengthy etymological discussion of the epic-poetic usages that connect the primordial gods of the sea, Metis, Eurynome, and especially Thetis, with the opening of pathways through the dark and trackless sea, a realm without markers (containing "trackless" Tartaros). In a Greek world bound by the sea, these goddesses are the ones who through their *mêtis*, open discernable paths, marked by the stars, within the boundless darkness of the sea. Seafarers (archetypically represented by the Argonauts) depend on them. If, as one tradition holds, Thetis is the creator of heaven, she shares the distinction with the celestial blacksmith in the tradition that saw heaven as a brazen bowl.[317] As will be seen in the following chapter, the connection of metallurgy with powers of the sea is a strong one in Mediterranean traditions. Hephaistos' strong

---

[314] Ibid, 142, italics mine.

[315] Ibid, 142.

[316] See: Detienne and Vernant, *Cunning Intelligence in Greek Culture and Society*, trans. Janet Lloyd (Atlantic Highlands: Humanities, 1978).

[317] Ibid, 142; Delcourt, 62-63.

connection with the element of water as well as that of earth is expressive of the archetypal power of the smith-god.

## The Return of Hephaistos

One of the signal qualities of Hephaistos is that he is both a master of binding and "master of a magic power of liberation."[318] Immediately following the delivery of the tricky golden throne and Hera's sudden entrapment in it is the dawning realization on Olympus that no god there is capable of freeing her. Her bonds are magical and only the maker can undo them (as is the case with all the Greek gods—only the god who does something can undo it). Zeus orders Hephaistos to be brought back to Olympus—for the first time since he was cast out by Hera? Or perhaps he has already been thrown off, again, this time by Zeus? From whence is he brought? No remaining tradition tells us. We are told that Zeus sends Ares to retrieve the tricky smith. It may be assumed that Ares is abrupt, even truculent in his summons. What the myth does say is that he receives a blast of forge fire for his pains and retreats, singed, to Olympus— where he is later shown, on the "François Krater" depicting Hephaistos' arrival, set apart from the other Olympians "kneeling and sullen-faced."[319]

Zeus next sends Dionysos, who succeeds where Ares failed, perhaps with flattering *logoi* and certainly with strong wine. The "Loosener," as Dionysos is known, succeeds for the moment in separating Hephaistos from his anger. Drunk, Hephaistos is loaded onto a mule (and is very occasionally represented draped on it as if he believed himself still at the banquet). He is conducted to Olympus in a procession led by horse-tailed Silenos (usually ithyphallic; often the mule is ithyphallic too, as in the "François Krater") and followed by Dionysos, who is crowned with vines and carries his *thyrsos*; or, Dionysos leads and Silenos follows the lame god's mule. Often they are accompanied by additional piping, prancing satyrs and nymphs. Hephaistos also is often shown crowned with vine leaves (a party favor from the drunken banquet?).

---

[318] Ibid, 84.
[319] Gantz, 76.

Hephaistos' return to Olympus is the scene by far most often depicted in vase paintings featuring the god. Writing in 1976, Frank Brommer reports that of all the then-known representations of Hephaistos, 280, or somewhat more than a third, were vase paintings. Of these, more than 180 represent the return of Hephaistos with Dionysos to Olympus. The vases, moreover, were made very widely, emerging from workshops not only in Athens, Corinth, and Sparta, but also Etruria, the Greek colonies in Italy, and in Asia Minor. As noted above, the story is mentioned by Alkaios (b. 620 BCE). On evidence of the earliest vases depicting the scene, Brommer asserts that the story must have appeared in some literary form at the latest by the beginning of the sixth-century BCE.[320] In one fifth-century red-figure painting, the figures of Hephaistos and Dionysos are shown in association with a satyr-player on a small stage, underscoring the undoubted rendering of the scene in a satyr-play. It is also thought, on evidence of two vases showing the throne room scene presented in the costume conventions and manner of stage comedy (one in London, in which figures labeled "Daidalos" and "Enyalios" are hero stand-ins for their respective associated gods, Hephaistos and Ares), that the story was in fact presented not only as a satyr play but in a comedy.[321] Gantz mentions a comedy by Epicharmos (550-460 BCE), called *Hephaistos* or *Komestai* ("Revelers"), and a satyr play by Achaios, *Hephaistos*, and that nothing is known of them but the titles.[322] From at least 150 vase paintings that certainly depict this scene, together with "countless" others that likely do, Brommer concludes that this story was in its time one of the best-known and favorite of stories in the Greek world.[323]

This scene remained widely popular for two centuries, but Hephaistos' drunken return to Olympus virtually disappeared from vase paintings after that time.[324] Fourth-century BCE coins from the Greek colony on the volcanic islands off Sicily show Hephaistos seated before the forge, his hammer in his lowered left hand and a goblet in his raised right hand, perhaps suggesting a memory of

---

[320] Brommer, 10.
[321] Brommer, 17.
[322] Gantz, 76.
[323] Brommer, 11.
[324] Ibid, 17.

drunkenness surviving in local tradition.[325] (I am wearing one of these coins around my neck as I write, a birthday gift from my husband.)

Another small class of vase paintings (Brommer lists nine), also dating from between the late sixth century and the early fourth century, depict Hephaistos banqueting with Dionysos or Hephaistos festively adorned and carrying his hammer, approaching a reclining Dionysos in the banquet room. The banquet scene is presumably where Hephaistos is made drunk before being packed onto a mule and led in procession by Dionysos, Silenos and prancing satyrs to Olympus, to free Hera from the sticky predicament into which wily Hephaistos had entrapped her. "Indeed," writes Brommer, "this is the story that unites both these gods and affirms the foreignness they share in common."[326] As to Hephaistos' foreignness, Brommer cites no fewer than seventeen prominent classical scholars writing between 1926 and 1975 who place the origin of Hephaistos in Asia Minor. However, Brommer finds that each in turn cites the same scholar, L. Malten, who, writing in 1916, based his conclusion on evidence of the significant number of Hephaistean coins found in the region— which Brommer has shown are datable six centuries more recently than Malten's original dating. Nor are there any other known representations (vases, sculptures, reliefs) in Asia Minor datable earlier than the Hellenistic period. In other words, the fame of Hephaistos (or more accurately, Vulcan, since the coins are Roman-era) spread to Asia Minor rather than the other way around. It simply cannot be said with certainty where Hephaistos' origins truly lie.[327]

As to the "foreignness" of Hephaistos and Dionysos, whatever their original origin—which in both cases may at some point be demonstrated unquestionably to be natively Greek—the story of Hephaistos' return places together two "outsider" gods who are newcomers to Olympus and legitimacy within the family of gods ruled by Zeus. It is difficult to say what the Dionysian connection may mean except that Hephaistos, who has seemed in certain ways so easy to characterize or categorize (by for example writers like

---

[325] Ibid, 69.
[326] Ibid, 17, translation mine.
[327] Brommer, 1-3.

Atchity) may truly be as manifold and complex as Dionysos. Certainly, the gods who possess *mêtis* do not stand still long enough to bear that kind of scrutiny—even lame Hephaistos.

What can be made of the evidence of these vase paintings? That Hephaistos should undertake an act of revenge on Hera as a faithless and abandoning mother is not surprising, and by Pausanias' time was regarded as an obvious idea. That Hera, depicted since Homer as the meddlesome, jealous, vain, and somewhat reckless wife of Zeus, should unwittingly fall into such a trap as that presented by the shining throne seems also to agree with her character as understood by the Greeks from at least the time of Homer. Yet the scene of Hera caught in her suspended throne is far, far less often represented. It is to be surmised that the satyr play of Epicharmos and perhaps as well the comedy of Achaios painted with the broadest strokes the glorious drunkenness of the god and the pranks of the ithyphallic satyrs—and as well the inevitably hugely endowed mule, sometimes rampant like the satyrs. (And is perhaps the sterile mule a statement on the suspect sexual adequacy of Hephaistos?) Artistically, perhaps, there would be both more interest and less technical difficulty for most artists in portraying the procession than in trying to figure out how to depict Hera's floating throne.

What I find compelling about the "Return of Hephaistos" is that the scene focuses on a different Hephaistos from the Homeric. He is again the butt of humor, but whereas in Homer's *Iliad* he draws laughter for his shuffling gait as he plays cupbearer pouring out ambrosia, and in the *Odyssey* when he chooses to make a public demonstration of his own cuckholding, the return to Olympus places him in a different context. This context suggests itself as the only instance where Hephaistos himself is bound, enchained as a middle link within the procession of Dionysos and satyroi fore and aft, the latter at once ithyphallic and curiously nurturing of the doubly impaired god. Further, two of the key Homeric scenes mention his being thrown *from* Olympus. In the Return, we are shown a transitional moment: Hephaistos is making his way *to* Olympus, his place there secured by virtue of the fact that he introduces into it archetypal qualities that were lacking.

On closer examination, Greek Hephaistos presents a complex and ambiguous image. He is a god—perhaps *the* god—of binding and unbinding. Hephaistos is not himself bound—unless his gait, which in its circular character may be thought to resemble the links of a chain, describes a bond that is itself paradigmatically representative of the binding power of *métis*. Hephaistos is the god of *technê* in the sense of instrumental making—Prometheus teaches, gods such as Hermes invent, Hephaistos *makes*. His divine connections encompass the elements of fire (forge fire, telluric-volcanic fire, sacrificial fire), earth (metallurgy), air (winds and bellows), and water (his undersea apprenticeship and *métis*). He is a god of ambiguous parentage and suspect sexuality. He is capable of persuasive speech and action born of self-consciousness. He is a member of the Olympic pantheon who has also felt the "mortal pain" of injury and whose permanently deformed condition is central to his identity and his archetype. In the following chapter, I will present images and traditions that will help to place Hephaistos more centrally within an archetypal field that is better understood by starting with additional Greek traditions that are not acknowledged in the Homeric presentation of the god, and then venturing into non-Greek mythologies, in order to see the full power of the mythic smith.

CHAPTER 4

# The Dwarf, the Bogey, and the Blacksmith

THE EXAMINATION in the previous chapter of the Greek myths and mythemes concerning Hephaistos shows the god to be an ambiguous and puzzling figure. Even the ordered and ordering depictions of the Zeusian cosmos depicted by Homer leave us with unresolved questions about the origins and meaning of the blacksmith god. Hints of Hephaistos' *mêtis*—his crooked gait, his animated creations, and his tricks like the binding of Hera in the golden throne and ensnaring of Ares and Aphrodite—reveal elements that are both magical and ungovernable. What are we to make of the laughter of the other Olympians on the rare occasions when he draws attention to himself? What are we to make of his "peacemaking" nature, as when he calms the company in the Olympian banquet hall and speaks encouragement to Thetis in his own mansion, set against his implacable rain of fire on the river god Xanthus when called by Hera onto the Trojan field? What are we to make of the excellence, magical luster, and even agility of his creations when compared to his own lack of beauty and his restricted motion? A backward look into the earliest Greek mythologies, and a sideways glance at non-Greek mythologies and folklore, will yield a wealth of imagery to apply to a more complete understanding of the outlines of the Hephaistos archetype. I will examine myths of what I will term the "chthonic phallic cohort," including the tradition of the dwarf in folklore and fairy tale, as well as Greek and non-Greek traditions that will serve to define the nature of the divine blacksmith. Finally, I will examine the specific nature of Hephaistean fire.

## The "Chthonic Phallic Cohort" of Deities

As has already been discussed, *mêtis* links its possessors with the earliest generations of Greek gods and mythic figures. These primordial figures include the mysterious male divinities who combine the knowledge of magic, art, healing, and technology, gifts that they receive from the Great Goddess. Ancestors, cousins, even "sons" of Hephaistos, their presence persists in myth and folklore, indicating the deep significance of the knowledge of the properties of the earth and the powers of making. The Kabeiroi, Daktyloi, and Telchines are often mentioned together, and are difficult to distinguish one from another. These figures have a phallic character and are associated with mysteries and magic. They are often seen as founders of peoples, representing their genealogical connection with divine ancestors or their heritage in connection with specific places on the earth.

The term *chthonic* from Greek *chthôn*, "earth," refers to that which is under the earth or connected with the underworld. Certain Greek deities were known as chthonic in connection with their underworld connections or functions, including Hades, Persephone, Demeter, Dionysos, Hecate, and Hermes. The term *chthonic phallic* is used by Erich Neumann to refer to the mythic aspect of the male psyche that is in thrall to the underworld, vegetative, animal nature of the female principle. Neumann builds on the work of Johann Jakob Bachofen, who examines mythology to explain the pre-Hellenic, "primitive" and "archaic" cultural stage of the matriarchal age, or age of "mother right," which "was overlaid or totally destroyed by the later development of the ancient world.[328] In archaic creation myth, "Visible creation proceeds from [woman's] womb, and it is only then that the sexes are divided into two, only then does the masculine form come into being.... The female is primary, the male is only what

---

[328] Bachofen, J. J., *Myth, Religion and Mother Right: Selected Writings of J. J. Bachofen*, trans. Ralph Mannheim, Bollingen Series 84 (Princeton: Princeton University Press, 1967), 69.

comes out of her." Thus, "the masculine principle is of second rank, subordinate to the feminine.... Herein is rooted that age-old conception of an immortal mother who unites herself with a mortal father. In a word, the woman first exists as a mother, and the man first exists as a son."[329] Further, in primitive myth, "Everywhere the material, feminine, natural principle has the advantage; it takes the masculine principle, which is secondary and subsists only in perishable form as an ever-changing epi-phenomenon, into its lap...."[330] Neumann adds that, "The young men whom the Mother selects for her lovers may impregnate her, they may even be fertility gods, but the fact remains that they are only phallic consorts of the Great Mother...." And, this is why "these youthful companion gods always appear in the form of dwarfs. The pygmies who were worshipped in Cyprus, Egypt, and Phoenicia—all territories of the Great Mother— display their character just like the Dioscuri, the Cabiri, and the Daktyls...."[331] The serpent so often connected with the Great Mother is a numinous "symbol of the fertilizing phallus." The Mother often appears in uroboric form, that of the snake swallowing its tail in an unbroken circle. She is:

> mistress of the earth, of the depths and the underworld, which is why the child who is still attached to her is a snake like herself. Both become humanized in time but retain the snake's head. Then the lines of development diverge. The fully-human end-figure, the human Madonna with the human child, has her forerunner in figures of the human mother with her companion snake in the form of a child or phallus, as well as in figures of the human child with the big snake.[332]

Erich Neumann describes two aspects of the masculine self. There is a "lower phallic variety" of masculinity that responds to the orgiastic demands of the Mother, who embodies nature's demand for fecundity, which involves a submerging of the masculine consciousness into the oceanic pull of orgasm and its attendant loss of consciousness. Opposed to this is a "higher" masculinity

---

[329] J. J. Bachofen, *Urreligion und antike Symbole*, 2:356-58, qtd. in Erich Neumann, *Origins and History of Consciousness,* Bollingen Series 47, trans. R. F. C. Hull (Princeton: Princeton University Press, 1970), 47-48.

[330] J. J. Bachofen, *Urreligion und antike Symbole*, 2:359, qtd. in Erich Neumann, 48.

[331] Erich Neumann, *Origins and History of Consciousness,* Bollingen Series 47, trans. R. F. C. Hull (Princeton: Princeton University Press, 1970), 48.

[332] Neumann, 48-49.

"correlated with light, the sun, the eye, and consciousness."[333] Eugene Monick, writing on Phallos, offers a critique to Erich Neumann's distinction between a "chthonic" masculinity which is "absorbed in the realm of sexuality, the domain of the Great Mother" and a "solar" masculinity representing a "higher" development "in a spiritual direction."[334] Though in Neumann's view this is a move necessary for the individuation of the male psyche, it is problematic for Monick. It presupposes that the chthonic is inherently feminine and "lower," and that the solar as evolved masculinity is superior "as if this were a natural law of the psyche." Thus, "a particularly virulent form of patriarchy is encouraged." [335] For Monick, the chthonic phallic is not to be thought of as the shadow of the solar phallic. Both solar and chthonic phallic have shadow sides.[336] The shadow of chthonic phallic is gross brutality and rape; the shadow of solar phallic is the narcissism of the institution. Both in their shadow aspects are productive of violence. What the male receives from the chthonic phallic, however, is identity with what Jung called "the source of life and libido, the creator and worker of miracles...."[337]

In examining what I term the "chthonic phallic cohort"—the mythic and legendary groups representing masculine powers anciently associated with art and technology—I use the term "chthonic" to express a set of mythic and religious associations that have underworld connections, that are indeed associated with powerful goddesses. These male figures present a complex set of qualities related to sexual generativity in regard to *making*. By this I mean their ability to create through means other than or in addition to specific genital sexual function. For example, as evidenced in three kinds of Hephaistean "children," Erichthonios, the son of Hephaistos, is engendered from Hephaistos' spilled seed, not from consummated sexual union in the usual sense; Hephaistos creates Pandora from earth, endowing her with the living qualities of "strength and voice;" he also fashions girls from gold who exhibit the same attributes of

[333] Ibid, 92.
[334] Eugene Monick, *Phallos: Sacred Image of the Masculine* (Toronto: Inner City Books, 1987), 57.
[335] Ibid, 60-61.
[336] Ibid, 93.
[337] Jung, *Symbols*, 146.

strength and voice. These creations represent a blurring of the distinctions between instrumental creativity and natural generativity.

The theme of a connection between metallurgy and magic characterizes several groups of figures who remain shadowy and devilishly difficult to define. The Kabeiroi, said to be named after Mount Kabeiros, the mountain of the Phrygian Great Mother, whose Samothracian cult was closely bound up with that of the ithyphallic Hermes, are sometimes identified with the Idaean Dactyls, named after Mount Ida, a mountain sacred to Rheia, mother of Zeus. The Daktyloi were supposed to have learned the blacksmith's art from Rheia.[338] Kerényi says that the Daktyloi are "dwarfish," and are associated with the chthonic phallic.[339] All are makers.

## Kabeiroi

The Kabeiroi are associated with mystery cults which had sanctuaries on Lemnos, and at Samothrace and Thebes. The specific connections of the Kabeiroi to Lemnos and Hephaistos are very strong. Several writers transmit the tradition that the Kabeiroi were offspring of Hephaistos and they were also known as the Hephaistoi. Lemnos is the center of Hephaistean cult; its capital city even today is called Hephaistia. According to Lemnian tradition, the Kabeiroi are sons of Hephaistos and Kabeiro, a daughter of the sea-god Proteus; alternately, the Kabeiroi are grandsons of Hephaistos and Kabeiro, tracing their origin to a son of Hephaistos and Kabeiro named Kadmilos (or Hermes).[340] Kerényi remarks that Strabo associates Kabeiro with Rhea, Demeter, Hecate or Aphrodite.[341] Kabeiro would thus be a great goddess of chthonic power, and Kabir is a term related to religious nomenclature.

The tradition of Hephaistos' mating with Kabeiro, daughter of Proteus, who like Thetis's father Nereus is one of the Old Men of the Sea, echoes the powerful oceanic motif in the Hephaistos mythos. The oceanic theme is also attested in the tradition invoking the Kabeiroi as *megaloi theoi*, "great gods," by seamen in times of

---

[338] Jung, *Symbols*, 127, 183.
[339] Carl Kerényi, *The Gods of the Greeks* (London: Thames and Hudson, 1951), 86.
[340] Gantz, 148.
[341] Kerényi, 87.

danger.[342] Indeed, part of the purpose of the mystery cult devoted to them is protection from death by drowning.[343] The marine motif also appears in the epithet *Karkinoi*, "crabs," for the Kabeiroi on Lemnos, where the crab was said to be specially revered, and specifically in connection with metallurgy: *karkinos* is also the word for the blacksmith's tongs. Detienne and Vernant point out that the crab was known by the Greeks both for its tong-like pincers and its oblique gait, citing the Greek proverb, "You will never make a crab walk straight."[344] Detienne and Vernant of course discern the image of the crooked gait of Hephaistos in the crab's sideways walk.

The connection of Hephaistos to Dionysos as evidenced in the popularity of the motif of the return to Olympus as a subject for vase painting was introduced in the previous chapter. The tradition of the Kabeiroi in Lemnos also links the two gods. In a lost play by Aeschylus, *Kabeiroi*, the Lemnian Kabeiroi receive the Argonauts on their island and "introduce themselves as prodigious wine drinkers," in connection with which Burkert notes that "wine vessels are the only characteristic group of finds from the Kabeiroi sanctuary on Lemnos." The burlesque element in Lemnian cult exemplified in the return of Hephaistos to Olympus, drunk, in the company of Dionysos and satyrs, is historically attested in a ritual dedication in connection with the Lemnian cult to the god who "jests by the way" (*parapaizonti*).[345] This may suggest a ritual origin for the laughter Hephaistos' actions occasion in Homer's *Iliad* and *Odyssey*.

The Kabeiroi have left more complex traces at their sanctuary in Thebes. Votive imagery shows a bearded god reclining to drink from a *kantharos*. In at least one example, this figure, whose attributes are those of Dionysos is instead labeled "Kabiros."[346] The excavated Theban sanctuary included rotundas with hearths, meant for sacred meals. A fairly common representation of Hephaistos shows him involved with reclining Dionysos in an intimate banquet scene.[347] Wine bowls found at Thebes are invariably in fragments, indicating

---

[342] Ibid, 86.
[343] Burkert, 284.
[344] Detienne and Vernant, 269-70.
[345] Burkert, 281.
[346] Harrison, 652.
[347] Brommer, 17.

that they were used once and intentionally broken, suggesting a Dionysian theme of ritual wine-drinking.[348]

Two additional facets of the cult associated with the Kabeiroi are evidenced in the mystery center in Thebes. One is the presence of the image of father and son. Votive dedications show Dionysos (or Kabiros) together with a boy, Pais. Playthings, mostly spinning tops (associated with Dionysos in Orphic mythology) dedicated to this boy survive among the ritual objects found, together with figurines of a boy wearing a *pileus*. Jung remarks that "the queer little pointed hat, the *pileus*…is peculiar to these mysterious gods [the Cabiri] and was thenceforward perpetuated as a secret mark of identification."[349] Jung reproduces a vase painting showing "Odysseus as a Cabiric dwarf" wearing a *pileus*.[350] The *pileus* became an attribute of Hephaistos in later images.

At Thebes, the tutelary great goddess was Demeter Kabeiraia. There, Prometheus is identified as one of the Kabeiroi and the goddess "instituted initiations there for Prometheus and his son, Aitnaios." Burkert says that in the Theban context Prometheus represents a connection to "guilds of smiths analogous to those of Lemnian Hephaistos." As one of the Kabeiroi, Prometheus would be reckoned one of the "sons" of Hephaistos. The father-son imagery indicates that male initiation rites were conducted at Thebes and probably at Lemnos, though few details are known of them, and nothing of the specific role of Hephaistos.

The Kabeiroi also appear at Samothrace. The names of gods involved in the mysteries at Samothrace were kept secret, but two bronze statues of ithyphallic Hermes flanked the entrance to the *anaktoron* (Holy of Holies), and a ram sacrifice ("a specialty of Hermes") was likely conducted as part of the probable initiation rites performed there. "In addition," notes Burkert, "there is a young god in the role of a servant who is mentioned frequently, Kasmilos or Kadmilos, translated as Hermes."[351] Kerényi speculates that the worship within the most holy shrine at Samothrace would have involved both a "small" and a "great" Kabeiros/Kadmilos, father and

---

[348] Burkert, 282.
[349] Jung, *Symbols,* 183.
[350] Ibid, 128, fig. 13.
[351] Burkert, 281-84.

son who were aspects of the same personage, who was in turn both husband and child of the Great Mother, a relation that is "often to be found in tales concerning our mysteries."[352] It is known that the initiate at Samothrace was asked by the priest to disclose the worst deed he had ever committed—"not so much to elicit a confession of sins as to establish complicity, thereby securing unbreakable solidarity." Following initiation, the initiates wore iron rings forever thereafter (might this be the origin of the ring worn by Prometheus binding him to Zeus?). Burkert mentions that the mysteries of Samothrace are first of all claimed to prevent drowning at sea and promote successful voyages. In different versions of the tales, the Argonauts, the mythical first seafarers of the Greeks, celebrated the rites of the Kabeiroi on Lemnos and underwent initiation at Samothrace.[353]

## Daktyloi, Kouretes, Korybantes

Kerényi reports on traditions that identify the Kabeiroi with the Daktyloi, who "had come westwards from Phrygia and whose magical practices had made the inhabitants of Samothrace the first converts to their secret cult."[354] The Daktyloi were known as "sons" of the Phrygian Great Mother. Details differ: Pherekydes speaks of two groups of Daktyloi, thirty-two sorcerers of the "left," who cast spells and twenty of the "right" who dissolve them. They are thus divinities of binding and unbinding. Sophocles explains their name as a derivative of their numbers, five each of male and female, like in number to the fingers of each hand (*daktyl* meaning "finger"). Another author reports that they are called Daktyloi "because meeting Rheia by chance on Ida they took her fingers in greeting."[355] Also described as dwarfish, the Daktyloi may be symbolically small in stature. They are named after their symbolic association with the fingers, hence "thumblings," as in the folk and fairytale tradition. The confusion of stories concerning their origin, however, suggests not size so much as instrumentality of a magical nature.[356] They were

---

[352] Kerényi, 87.
[353] Burkert, 283-84.
[354] Kerényi, 86.
[355] Gantz, 149.
[356] Thompson, Folktale, 248.

widely known to many classical authors as sorcerers and the first forgers of iron. According to Diodoros they received the knowledge of working iron from the "mother of the gods," Rheia. The Daktyloi are thus representative of the magical and instrumental powers and symbolism of the hand, and specifically in connection with metallurgy and forging. The nature of the left and right-handed magic of the Daktyloi may provide a clue to Hephaistos' epithet *amphigueeis* "ambidextrous." Were certain of Hephaistos' creations of the "left" hand—the binding throne of Hera and the golden net—and some of the "right," like the shield made to ensure the fame of Achilles on the Trojan battlefield.

Mythically, like the Kabeiroi, the Daktyloi are also known as foster-parents. The care of the infant Zeus is given into their hands by his mother Rheia. Pausanias says the Daktyloi were the same as the Kouretes, known as playful "dancers" or "flute-players" who were children of the daughters of Doros and siblings of the mountain Nymphai and the Satyroi, and linked to the Daktyloi through the foster-parenting mytheme.[357] The cave of the Kouretes in Crete is the birthplace of Zeus. It was there that they hid Zeus from Kronos, drowning the sound of his cries by rattling their armor. To add to the confusion between these groups of divinities connected to magic and making and ultimately by their affinity with metal to the Hephaistean complex, the Kouretes are also either companions to or identified with yet another group, the Korybantes. In another version of the infancy of Zeus in Crete, the Korybantes invent the drum, which they give to Rheia—its noisy function may be similar to the armor-rattling of the Kouretes.[358] The phallic nature of the Kouretes is further affirmed through a Macedonian tradition reported by Kerényi of the Korybantes that they "carried the basket of mysteries, containing a phallus, the male member of Dionysos, to the country of the Etruscans."[359]

---

[357] Gantz, 147.
[358] Ibid, 148.
[359] Kerényi, 86-87.

## Telchines and Keres

Yet another associated group is the Telchines. Like the other figures under discussion, the Telchines are metal-workers, forging bronze and iron. Diodoros cites them as the first inhabitants of Rhodes.[360] Like the Daktyloi, Kouretes and Korybantes, the Telchines too are foster-parents. Strabo credits them with fostering Zeus in Crete at the behest of Rhea. Diodoros says that Rheia gave Poseidon into their care. Diodoros also reports that they were artists who made the first images of the gods.[361] This connection is Daidalic, underscored by their legendary presence at Crete, along with the Kouretes and the Daktyloi, who, according to a Hesiodic fragment, discovered the working of iron there.[362]

Like the Kabeiroi and Daktyloi, the Telchines are masters of magical technologies. They are notable sorcerers and are especially associated with the dark arts. Harrison cites a passage from Stesichorus describing the "ancient tribe" of the Cretan Kouretes, "also called Thelgines," who were sorcerers and magicians:

> Of these there were two sorts: one sort craftsmen and skilled in handiwork, the other sort pernicious to all good things; these last were of fierce nature and were fabled to be the origins of squalls of wind, and they had a cup in which they used to brew magic potions from roots. They (i.e. the former sort) invented statuary and discovered metals, and they were amphibious and of strange varieties of shape, some were like demons, some like men, some like fishes, some like serpents; and the story went that some had no hands, some no feet, and some had webs between their fingers like geese. And they say that they were blue-eyed and black-tailed.[363]

The Telchines' magical and skillful connection with the sea is also strong. The Telchines are variously known in Greek tradition as "water wardens," "sons of the sea," or "children of the sea."[364] The testimony of Stesichorus, above, recalls the shape-shifting capabilities of the Okeanids, capabilities that are emblematic of *mêtis*.

---

[360] Gantz, 149.
[361] Kerényi, 88.
[362] Gantz, 148.
[363] Harrison, 172.
[364] Morris, 92.

The Telchines, too, are children of the earliest generation of divinities. They are said to be either children of Pontos and Gaia (Pontos, Ocean, is both the offspring and mate of Gaia), made without sexual congress; or the offspring of Tartaros, the personification of the lowest part of the cosmos, and Nemesis, daughter of Nyx, or Night.[365] Their connections with the earliest generations of divinities is also shown in the tradition that equates the Telchines with the Keres.[366] The Keres are produced by Nyx, who, like Gaia, emerges directly out of Chaos. The Keres are spirits, evil sprites who are bringers of disease or vicious, frightful demons who fasten sharp teeth and claws on the dead, dragging them to the underworld.[367] Blindness is caused by the Keres, "hence the expression 'casting a black Ker on their eyes.'" Harrison connects the frightening character of the Keres with that of the Erinyes, noting that "Blindness and madness, blindness of body and spirit, are scarcely distinguished, as in the blindness of Oedipus; both come of the Keres-Erinyes."[368] In Ovid's *Metamorphoses*, Zeus "submerges" the Telchines—another watery connection—"because their eyes harm all things with their gaze."[369]

It is the Telchines who forge the sickle used by Kronos to castrate his father Ouranos when out of jealousy Ouranos prevents Gaia from bearing the Titans, who are thus bound in her womb unable to be born.[370] In her helplessness to overcome Ouranos, Gaia appeals to her unborn children. Timothy Gantz makes the observation that the castration of their father seems to have been accomplished while the Titans were yet in their mother's birth canal.[371] The Telchines as representative of the chthonic phallic cohort align with the Mother through instrumentality, *making*: they make the implement used to work her will in ridding herself of the phallus of Ouranos and allowing her to birth her children (and indirectly, for Aphrodite to be generated when the severed phallus falls into the sea). This image

---

[365] Gantz, 3, 9-10, 149.
[366] Harrison, 171.
[367] Harrison, 170, 175.
[368] Ibid, 167.
[369] Gantz, 149.
[370] Ibid, 149.
[371] Ibid, 10.

must ultimately relate to the incestuousness of the relationship of the chthonic phallic cohort with the Mother. This incestuousness is not necessarily specifically sexual in terms of the son mating with the Mother—though an enactment of such a union may well have figured in rituals celebrated in Samothrace or elsewhere—but it is collaborative in nature.

## Cyclops

Harrison cites the relationship of the Cyclops to the Daktyloi, as indeed generally to the "craftsman [who] is regarded as an uncanny bogey himself, cunning over-much, often deformed, and so he is good to frighten other bogeys."[372] At his forge inside Etna, Hephaistos is assisted by Cyclops. The Cyclops are primordial figures in Greek myth, according to Hesiod's *Theogony*, sons of Gaia born after the Titans (and thus older by two generations of gods than Olympian Hephaistos). Originally, they are "three in number and like to the other gods in all things save for the single round eye in their foreheads."[373] They are connected with lightning and thunder and forge the thunderbolts of Zeus—including the one with which Zeus kills Asklepios, the son of Apollo, for daring to use his healing gifts to rescue several humans from death. Unlike humankind, the Cyclops are divinities whom we should expect to be immortal, although according to a "Hesiodic" fragment (from the *Ehoiai*) Apollo punishes the Cyclops for Asklepios' death by killing them. According to another variant in a fragment of Pindar, Zeus kills them so they cannot forge weapons for anyone else.[374] Still another variant (in Virgil's *Eclogues*), tells of "hostility between the Telchines and Apollon, as a result of which the younger god destroyed the older ones."[375] The source of this hostility is not told, but the Telchines, too, were credited with forging the thunderbolts of Zeus as well as the trident of Poseidon.[376] It is tempting to surmise the reason lies in a confusion with the tale of the Cyclops, especially in view of the frequent inter-identification of the chthonic phallic cohorts.

---

[372] Harrison, 190.
[373] Gantz, 10.
[374] Ibid, 13.
[375] Kerényi, 88.
[376] Gantz, 149.

Polyphemos, the one-eyed Cyclops who is blinded by Odysseus in Book 9 of Homer's *Odyssey*, is said to be a son of Poseidon. He and his neighbor Cyclops are pastoralists—they raise sheep but are brutes, ignorant of the skills of husbandry. They have no *technê*. They are, however, related to the people of Alcinous, the most excellent of shipbuilders and clients of Hephaistos, who creates the golden guard-dogs and other features of Alcinous' palace (which we learn in Book 7 of the *Odyssey*). There would seem to be no other connection between them and the primordial Cyclops, except for the tradition in Ovid noted by David L. Miller that the Cyclops were originally *smiths* and that, according to Apollodorus, "they were the armor-makers for Zeus until Apollo reduced them to shepherd-status in a dispute over Asklepios' ability to eliminate death, depriving Hades of his part in the nature of things."[377] The mytheme of the smiths' punishment is echoed in the tales of the Telchines.

## *Deformity as Forge-Mark of the Chthonic Phallic Cohort*

One of the signal features of this complex of inter-related images is some physical characteristic or marking, often a "disability," which separates these mythic figures from the rest of gods and humankind. Although the earliest generations of gods include many monstrous forms, the peculiarities of the chthonic phallic cohort are specifically connected with their powers as metalworkers and magicians. Miller observes that,

> The mark of Cain…is the mark of one who works the forge. It is a mark in the middle of the forehead, not as a punishment for moral disobedience, but as signifying the fire (remember *agni*, 'ignite') of the forge. It is like the solar eye of the Cyclops, but blackened by occupation, hence *sol niger*, 'the black sun.'[378]

The Greeks viewed the operations of the forge, kiln, and oven as magical and dangerous. The symbolic mark borne by the forge-worker, sometimes imagined in connection with the eye, recalls the fact already noted that the Daktyloi and Telchines were specifically associated with the evil eye (and indeed, Miller notes that shamanic

---

[377] Miller, *Christs*, 39.
[378] Ibid, 40-41.

initiation often included the initiate's hallucinatory experience of his or her eyes being torn out). Harrison describes the small figures and masks that served as apotropaic (i.e., serving to avert ill-will) kiln and forge bogeys, and indeed such could be found near the chimney in most Athenian homes. Harrison describes ovens decorated with the miniature masks of a satyr with pointed ears, compressed features and hair standing on end, and a bearded figure wearing a *pileus* (the egg-shaped/phallic-shaped "Phrygian cap") and surrounded by symbols representing lightning bolts.[379] These objects "are the equivalent of the Latin *turpiculum*: an object that is obscene but not shocking; rather it amuses and provokes the spark of laughter that breaks a curse."[380] Laughter and superstitious fear are evoked in the same figure. The ancient gesture of the "fig," mimicking phallus and scrotum, provides very much the same function. The laughter Hephaistos draws from the other gods in the Olympian banquet hall and standing beside his own nuptial bed which holds the adulterous Ares and Aphrodite captive arises similarly from the company being presented with the antics of the ugly, limping god. In both cases, the laughter shatters a charged, even potentially violent, situation whose outcome is at the outset unpredictable.

Dwarfism is also characteristic of the descriptions of the chthonic phallic cohort. Hephaistos himself is described as dwarfish. Herodotus reports that the image of Hephaistos in his temple at Memphis (the site of a Kabeiric sanctuary) resembles the figure of the Egyptian creator-artisan god Ptah, who is represented as a dwarf: pygmy-like, crooked-legged and large headed. Herodotus says that the images in the temple there of the Kabeiroi are like that of Hephaistos and were said to be his sons. The seagoing Phoenicians carried a Ptah figure in the prows of their triremes, underscoring the religious implications of the maritime migrations of the itinerant Bronze-Age forge-workers of the Aegean and Mediterranean.[381]

---

[379] Harrison, 190-91.

[380] Delcourt, 113, translation mine.

[381] Herodotus, *The Histories*, 3.37.2-3, trans. A. D. Godley (Cambridge. Harvard University Press, 1920), Perseus Digital Library Project, ed. Gregory R. Crane, 2004, Tufts University, accessed September 9, 2004, http://www.perseus.tufts.edu/cgi-bin/ptext?doc=Perseus%3Atext%3A1999.01.0126.

Historically, the ubiquitous figures introduced above represent the spread of metalworking eastward and southward through the Mediterranean in the Bronze Age and the secrets of powerful technologies carried by itinerant *technitēs* (possessors of skills) and *dēmiurgoi* (recognized and honored practitioners of valued skills such as medicine and shipbuilding) who traveled from place to place. Dwarfish, associated with magic, mysteries and the Great Goddess (Rheia, Demeter, Aphrodite, etc.), the chthonic phallic cohort of divinities are related to the earliest, most primordial generations of gods and are siblings and cousins of divinities of the earth and elements (*nymphoi, satyroi*). The dwarf godlings and races are weapons-makers, but are also child-fosterers (of Zeus and Poseidon) as well as healers. Asklepios the healer is said to have had a Kabeiric helper, Telesphoros.[382] Ultimately connected to all of the gods in one way or another, they are most closely related to Hermes and Dionysos through their mystery cults, involving "the opposition of old and young, male and female, sexuality and death."[383] And, these figures, whom Morris calls the "mischievous and mobile spirits of metal," are of course uniquely and centrally related to Hephaistos through their knowledge of iron and bronze working.[384]

## *The Image of the Dwarf in Folklore and Fairy Tale*

More than in perhaps any other imaginal figure, traces of the mythic and psychic significance of the chthonic phallic cohort can be seen in the persistent image of the dwarf in folklore and fairytale. Jung remarks on the phallic significance of the plaster gnome lurking innocently in Swiss gardens, wearing "the traditional headgear of our infantile chthonic gods today, the pixies and goblins"—a revenant of very ancient and uncanny mythic significance.[385] The phallic significance of the dwarf is connected with generative creativity but not in the way that human male-female genital sexuality issues in human offspring.

---

[382] Daniel C. Noel, "Veiled Kabir: C. G. Jung's Phallic Self-Image." *Spring*, 1974), 228.
[383] Burkert, 285.
[384] Morris, 88.
[385] Jung, *Symbols*, 183.

Dwarfishness is one of the attributes of the chthonic phallic cohort, and represents the "mark" of difference expressed in deformity—that is, recognizable divergence from the human norm, not merely in appearance but in being. This difference has traditionally been imagined in relation to a category that is profoundly significant to the human imaginal and corporeal experience: sexuality. That Hephaistos is sexually suspect is not an attribution of sexual impotence but instead an expression of the mythic imperative that is imaginally embodied in the chthonic phallic cohort. Bandy-legged dwarfs and their kind in folk- and fairytale usually do not marry. In Germanic and Nordic myth, "Dwarfs give rise to living beings only through the magic of their craft."[386] The dwarfish divinities predate the patriarchy and mythically partake in the earliest emergence of forms from the chaotic prequel to creation. It is their purpose to mold the partly formed clay, stone, and mineral of the universe into its myriad forms.

The mysterious, magical, artisanal and sexual characteristics of dwarves are more fundamental to their mythic nature than their size. There are conflicting reports on the size of the dwarf, and there is a rather disconcerting confusion as to whether dwarves are in fact fingerlings or giants, or both—Alice-like, very large and very small. Jung comments that this "liking for diminutives on one hand and superlatives on the other" is "connected with the queer uncertainty of spatial and temporal relations in the unconscious," and he notes that Goethe says of the Kabeiroi that they are "'little in length / mighty in strength'" in magic, healing, and artisanry.[387] Motz points out that the word *dwarf* is related both to the "Old Indian *dhvara*, a demonic being, and to an Indo-European root *dhver- with the meaning of 'damage.'"[388] It is more likely that the size of the chthonic phallic cohorts, whether abnormally large or small, taken together with images of ugliness, lameness, and deformity, represent the image of a race implacably *other*.[389] The dwarf is perhaps more prevalent in Scandinavian and German mythology and folklore than that of any

---

[386] Motz, , 91.

[387] Jung, *Symbols*, 408.

[388] Motz, 117.

[389] Jung, *Symbols*, 183.

other region, though dwarf tales occur in Egypt and the Middle East, India, Asia, Polynesia and Africa.[390]

According to the Icelandic *skald* (bard) Snorri, composer of the *Edda*, the abode of the Dark or Black Elves, who are dwarves, is "in the earth, and they are darker than pitch."[391] Of their creation, the *Edda* speaks of an *ur*-being, the giant Ymir, who was made of a kind of yeast that brewed itself out of ice melted in the origin chaos. Frost-giants and other beings were generated from him. His great-grandsons Odin, Vili and Ve killed him, and from his flesh was formed the earth, the sea from his blood, the rocks from his bones, and gravel and stones from his teeth. The dwarves came alive in the earth of Ymir's dismembered body, or underneath it, "like maggots in the flesh." Remembering their presence, the gods gave them human form and understanding. They still inhabit the earth and the stones.[392]

Thus, the dwarves are among the oldest-created—if not "older than dirt," then of an age with it. They are not born of sexual union. Indeed in Norse myth their creation is almost an afterthought. Dwarves, like the first men, are born out of dirt. The phallic dwarves replenish their own generations by fashioning them of clay or mud. Jung observes that "The Latin *lutum*, which really means 'mud,' also had the metaphorical meaning of 'filth,'" i.e. excrement.[393] In similar imagery, the creation myths of the Native North Americans often concern a succession of animals diving into the primeval flood waters to secure in their claws bits of mud which magically expand to form the earth.[394] According to Alan Dundes, this mud must be equated with feces and anality, thereby, he argues, demonstrating the origin of the "earth-diver" myth in the male's envy of the female's procreative power in creating life out of her body. By necessity, male creation out of the body is anal.[395] (Jung also mentions the creative aspect of

---

[390] Stith Thompson, *Motif-Index of Folk-Literature* (Bloomington: Indiana University Press, 1966), 103.

[391] J. A. MacCulloch, *The Celtic and Scandinavian Religions* (Westport: Greenwood, 1973), 120.

[392] MacCulloch, 154.

[393] Jung, *Symbols*, 279.

[394] Alan Dundes, "Earth-Diver: Creation of the Mythopoetic Male," in *Sacred Narrative: Readings in the Theory of Myth*, ed. Alan Dundes, (Berkeley: University of California Press), 1984), 277.

[395] Dundes, 279.

micturation in the Vedas;[396] and Gaston Bachelard cites the medieval writer Theophilus, who "recommended that iron be tempered 'in the urine of a he-goat or a red-haired child'"[397]).

Dundes cites Géza Róheim's theory that primitive myth recalls the "basic" dream: the dream of falling into a hole that every dreamer seems to experience, "characterized by a 'double vector' movement consisting both of a regression to the womb and the idea of the body as penis entering the vagina."[398] Róheim regards the earth-diver myth as an erection fantasy of diving into the primeval waters of the womb of the Mother. In light of the incestuous connection of the chthonic phallic cohort with the Great Goddess expressed in the Kabeiric mysteries, I would suggest that the phallic fantasy attributable to the earth-diver must in the case of the earth-generated dwarf be seen more precisely as evidence of the deep connection between Mother Earth with her sons, in a sacred sexuality that reveres both the womb and phallus as numinous. The image of the dwarf-deities is a representation of the specifically magical and supremely effective powers of the chthonic phallic over matter, creative powers that are effected in collaboration with the feminine rather than in differentiation or opposition to it, as is required by Neumann's notion of the solar phallic. The mytheme of the (male) offspring (Kronos) collaborating with the mother (Gaia) from within the womb in association with male powers (the Telchines) also provides a mythic example of Jung's observation of the alchemical principle that as the "female lies hidden in the male, so the male lies hidden in the female."[399] Indeed, the smith is an alchemist.

The dwarf's abode is under the mountain, in the mine. Eliade mentions the "primitive conception of mineral embryology," in which stones and ores grow in the womb of the Earth and engender precious stones.[400] For example, an Indian treatise on precious stones "distinguishes diamond from crystal by a difference in age expressed

---

[396] Jung, *Symbols*, 322.

[397] Gaston Bachelard, *Earth and Reveries of Will*, trans. Kenneth Haltman (Dallas: The Dallas Institute, 2002), 109.

[398] Dundes, 285.

[399] Jung, *Symbols*, 324.

[400] Mircea Eliade, *The Forge and the Crucible: The Origins and Structures of Alchemy*, 2nd. ed. (Chicago: University of Chicago Press, 1956), 49.

in embryonic terms: the diamond is *pakke*, i.e., 'ripe', while the crystal is *kaccha*, 'not ripe', 'green', insufficiently developed." European natural science at the dawn of the age of Enlightenment preserved a similar tradition that "The ruby, in particular, gradually takes its birth in the ore-bearing earth…. Just as the infant is fed on blood in the belly of its mother so is the ruby formed and fed." Pliny reported that mines "were reborn" after being closed up and allowed to rest after exploitation. A similar belief seems to have been shared by African metallurgists. Like vegetal life, "Ores 'grow' and 'ripen.'"[401] This is the elemental world that is the dwarf's domain.

In the myths of creation from earth/clay/mud, Dundes also traces the linkage of excrement with gold and money ("filthy lucre"), which inevitably recalls the dwarf metal-worker and earth spirit with knowledge of the precious metals.[402] While the dwarf commands riches, he values them differently from humanity. This connection can be seen in the familiar Rumpelstiltskin tale, in which the daughter of a poor miller is obliged to spin straw into gold. A "little man" (represented by the illustrators of the tale as a dwarfish figure) appears to her in her distress, and takes payment from her to work the magic of spinning straw into gold. Seeing the pile of gold, the king decides he cannot find a richer woman anywhere, and promises to marry her if she will spin more straw into gold. The young woman has nothing more to offer the dwarf, so she offers him her first born once she is queen, a bargain he gleefully accepts. When she duly marries and has a child, the dwarf returns and demands payment unless she can guess his name. The ending of the story of course is that the young woman magically learns the name, and Rumpelstiltskin tears himself asunder from frustration. The little man shows all the characteristics of the mythic chthonic dwarf. He has magical knowledge of a technology involved with labor/making (spinning) and with gold as the most precious of earth's products. Yet gold is not what is of most value to him, in contrast to the greedy king. He first takes as payment a necklace and ring, presumably adding these to the oft-mentioned dwarf treasuries, but what he wants is the child. Dwarves cannot procreate as humans can, and to obtain a

---

[401] Ibid, 44-46.
[402] Ibid, 285.

child, the little man must resort to bargaining with the human. Why he wants the child the tale does not tell, but Rumpelstiltskin belongs with the dwarf folktale type category, 451, which identifies the magical and transformative and nurturing abilities of dwarfs, for example tale-type F451.3.3.1, "Dwarfs turn peas into gold pieces," and tale-type F451.6.1.1, "Dwarf as godfather."[403]

Although they are of the earliest generations of created beings, the dwarves have "no share in the ruling of the elements of nature or in the shaping of the world [but they] are vital partners in its maintenance and preservation."[404] And it is to the dwarf creator-divinities and their unique, magical powers of making that the gods must resort, for the dwarves have the power to create what the gods cannot. The dwarves of Nordic myth created hair of gold for Thor's wife Sif to replace the hair the trickster-god Loki cut off. They also created the god Frey's boar, which could run through water and gave off bright light. Thor's miraculous hammer always struck true and returned to his hand when he threw it. Odin's similarly reliable spear Gungnir and Högni's unfailingly deadly sword Dainslef were also made by the dwarves. They made Skidbladnir, the ship of the gods which the breeze always favored and Frey could fold up and put in his pocket; and they made Odin's miraculously multiplying ring Draupnir.[405] The smooth, unbreakable fetter to restrain the monstrous Fenris-wolf was also made by the dwarves—out of things which had no existence: a woman's beard, the noise of a cat walking, fish-breath, bird-spit.[406]

Dwarves' association with smithing is not restricted to metalwork. The connection between *poiēsis* and all creative making, including poetry, has already been noted. Jung remarks that the Daktyloi, "to whom the mother of the gods had taught the blacksmith's art" are also associated with word-smithing. Stith Thompson lists an Irish folkloric motif of the first poetry being written in imitation of the rhythm of anvil-blows.[407] Jung notes that the dwarfish Daktyloi are said to be the teachers of Orpheus—meaning that Orpheus is among

---

[403] Thompson, *Motif-Index*, 103ff.
[404] Motz, 91.
[405] MacCulloch, 121.
[406] Motz, 93.
[407] Thompson, Motif-Index, A1464.1.1.

those initiated into their mysteries—and that they are the inventors of musical rhythms, "Hence the dactylic meter in poetry."[408] Poetry of course is far more than simple word-smithing, and the close association of the dwarf smith with poetry is rather an expression of the dwarves' magic and their connection with primordial creation. Moreover, myth tells us that poetic making is not for the faint of heart. In Nordic mythology, the dwarves kill the "man of wisdom" shaped out of the gods' own spittle and brew his blood into mead that could create a *skald* (bard) of the man who drank it. "Hence poetry was called 'the dwarf's drink.'"[409] Powerful song is an element of many myths concerning the divine smith. Eliade offers many examples, noting that the Phoenician smith-god Chusôr "invented the art of 'good speech' and that of composing chants and incantations." In Ugaritic texts the chanters are called Kôtarât," a title having the same root as the name of the smith-god Kothar-wa-Hasis. In Arabic, "*q-y-n*, 'to forge,' 'to be a smith,' is related to the Hebrew, Syriac and Ethiopian terms denoting the act of 'singing,' 'intoning a funeral lament.'" Similarly, "Odin and his priests were called 'forgers of songs'" and the same theme is found among the Turco-Tartars and Mongols.[410] Eliade also observes that "The Sanskrit *taksh*, meaning 'to create,' is employed to express the composition of the Rig Veda songs."[411]

The *Rg Veda* speaks of Brahmanaspati welding this world together "like a blacksmith" from the broken fragments of the dismembered corpse of Purusha, the "First Man" celebrated in Vedic hymns, who in effect explodes himself into fragments which become the countless manifestations of the world as we know it.[412] Jung refers to the self-sacrificing Purusha as a dwarf, a theme that survives in the fairytale motif of Rumpelstiltkin tearing himself apart.[413] Indeed, Eliade remarks that "creation is effected by immolation or

[408] Jung, *Symbols*, 183, n. 15; Kerényi, 86.
[409] MacCulloch, 121-22.
[410] Eliade, *Forge*, 98-99.
[411] Ibid, 98.
[412] Jung, *Symbols*, 178; William K. Mahony, *The Artful Universe: An Introduction to the Vedic Religious Imagination* (Albany: State University of New York Press, 1998), 27.
[413] Jung, *Symbols*, 182.

self-immolation."[414] In many traditions, sacrifice assures the efficacy of the forge. Such was the purpose of the practice of burying an embryo beneath the forge to assure its effectiveness.[415]

The activity of the world-smith Brahmanaspati embodies the "prayer-word" which, manifested by the goddess of sacred sound and the creative energies channeled through it (Vac), forms the different poetic meters, giving order to space.[416] The Vedic poet-priests, receiving divine inspiration, used the meters to construct the hymns they sang in sacrificial ritual to re-make the world at every sunrise just as Brahmanaspati did at the beginning of the world. In connection with the power of sacred sound and poetry to manifest the physical world, it makes poetic sense that the Norse myths include the tradition that the primordial gods assigned to the dwarves the task of standing at the four quarters as pillars to hold up the fabric of the universe.[417]

As pillars of the sky, the dwarves literally support a universe they have not created. However, they create the thunderbolts of the sky gods. For this reason, Motz remarks that we must see them as servants of the ruling dynasty.[418] Yet, they are primarily aligned with Earth, and it is from the Mother they have received their skills and the raw material which they transform through craft. The chthonic phallic dwarf has labored underground for millennia and with him the nurturing aspect of the male that works in collaboration with the life-bearing female. Our fairy tales reflect a human ambivalence of the greatest depth. Can one source of their continuing fascination be a living reminder that we must mind the balance? The primordial, perennial, ubiquitous dwarves remind us of the libidinous fire of desire and energy that stimulated the bursting of creation of the first body, Purusha. They also remind us of the sacredness of life and that its value can never be counted in gold.

---

[414] Eliade, *Forge*, 31.

[415] Forbes, 75.

[416] Jung, *Symbols*, 55-57; William K. Mahony, *The Artful Universe: An Introduction to the Vedic Religious Imagination* (Albany: State University of New York Press, 1998), 212-13.

[417] MacCulloch, 122.

[418] Motz, 91.

## The Blacksmith Lineage

The dwarf of fairytale and folklore holds several strands of the lineage of the archetypal blacksmith. The term *lineage* can be understood in two ways. First, in the sense that the divine blacksmiths of various mythic traditions are closely related to each other through specific mythemes. Second, that the magical power of the deity is also invested in the human smith as a son, who practices a sacred craft inherited via initiation. In many cultures, the professional status of the smith is attested through the transmission of genealogies demonstrating the smith's descent from his (or, rarely, her) divine patron. Eliade mentions the divine descent attributed to the smith in Indonesia and Africa and remarks that the knowledge and ability to recite the genealogies proving the divine descent of the smith, like that of other cultural heroes, is "the beginnings, as it were, of epic poems." He remarks that "This relationship between shamans, heroes and smiths is strongly supported in the epic poetry of central Asia." Indeed, "Certain aspects of the kinship between the craft of the smith and epic poetry are perceptible even today in the Near East and Eastern Europe where smiths and Tzigane [Roma or "gypsy"] tinkers are usually bards, singers or genealogists."[419] The relation between the poet and smith becomes even clearer in light of one of the purposes Eliade cites for epic poetry, namely to transmit genealogies of heroes.

Attribution of magical power to iron preceded the ability of humankind to forge it; it was initially inspired by iron falling from the sky in meteoritic form. Eliade cites the oldest word for iron, the Sumerian *an.bar*, which is composed of the pictograms for "sky" and "fire."[420] Among other words, the Greek word *sideros*, "related to *sidus, -eris*, meaning 'star,' and the Lithuanian *svidu*, 'to shine,' and *svideti*, 'shining'" may also attest the imaginal origins of iron as celestial.[421] Meteorites were worshipped in many places as images of deity:

---

[419] Eliade, *Forge*, 88.
[420] Ibid, 22.
[421] Ibid, 23.

The Palladium of Troy was supposed to have dropped from heaven, and ancient writers saw it as the statue of the goddess Athena. A celestial origin was also accorded to the statue of Artemis at Ephesus.... The meteorite at Pessinus in Phrygia was venerated as the image of Cybele.... A block of hard stone, the most ancient representation of Eros, stood side by side with Praxiteles' sculptured image of the god.... Other examples could easily be found, the most famous being the Ka'aba in Mecca. It is noteworthy that a certain number of meteorites are associated with goddesses....[422]

(It has been suggested that the Black Madonna image may also have meteoritic origins.)

Sidereal iron and by association terrestrial iron is linked to lightning and the thunderbolts and weapons of the ruler-gods forged by divine smiths. These weapons figure in many creation myths about the battles of gods over monstrous and chaotic forces. Zeus destroyed Typhon by hurling a thunderbolt forged in the smithy of the primordial Cyclops (many other versions attribute the forging of Zeus's thunderbolts to Hephaistos). The dragon Uritra in early Indian myth was slain by a brazen club. Eliade asserts that when the primordial storm-gods "strike the earth with 'thunder-stones'...the storm is the signal for the heaven-earth hierogamy. When striking their anvils smiths imitate the primordial gesture of the strong god; they are in effect his accessories."

According to Eliade, the mythology of metallurgy is a far more recent technology than those of food-gathering and small-game hunting, representing a shift in religious conception from "a *creatio ex nihilo,*' accomplished by a supreme heavenly deity [who becomes] overshadowed...by the idea of creation by hierogamy...." The "strong God" fertilizes the terrestrial Great Mother, and "we pass from the idea of *creation* to that of *procreation.*" Creation through the self-immolating primordial giant is superseded by sexualized creation, but retains its prototypical rootedness in sacrifice, so that "the stage [is] reached where creation or fabrication will be inconceivable without previous sacrifice."[423] That Hephaistos, as god of fire, is also the god of the sacrifice offered to and devoured by ritual fire, would seem to support this.[424] However, though a more

---

[422] Ibid, 20.
[423] Eliade, *Forge*, 30-31.
[424] Farnell, 374.

thorough examining of such a question is beyond the scope of the present work, I suggest that the blacksmith archetype should by no means be seen only as an image that serves the "strong god" through technological efficacy, but rather as a multivalent image that preserves as well a memory of *creatio ex nihilo* through its connection with the primordial, elemental, chthonic powers and divinities. The ambivalence and ambiguity of the smith image is not reducible, and the true genealogy of the smith lineage is older than the idea of paternally reckoned descent.

Bachelard notes the "symbiosis between flesh and metal" and that the role of the smith-gods in human and animal creation is attested in many mythologies.[425] The West African smith-god Ogun refines newly-created humans, by "adding lineage marks on the face and tattoos on the body, performing circumcision and other such surgery necessary to keep an individual in good health and make him or her socially acceptable in Yoruba society."[426] Talos, the man of bronze, is in one variant a remnant of Hesiod's race of bronze. In other variants the creation of Hephaistos bleeds molten ichor. Elsewhere, Talos is recorded as a progenitor in a line of descent that includes Hephaistos.[427] In Dogon myth, humans were formed and then broken at the knees and elbows by a divine blacksmith's hammer.[428] In the *Kalevala*, the divine blacksmith Illmarinen makes from iron, steel, gold, copper, and silver an eagle, an ewe, a stallion, and a maiden.[429]

In very early traditions, the smith is known as the wise one, the clever one. Like Daidalos, who is most likely a personification of the excellent and fateful ("*daidalic*") character of the objects made by the divine smiths, the ancient Ugaritic smith-god Kothar-wa-Hasis, the "Clever One" who features in the Baal epics of the early Bronze Age (late third millennium BCE) may be a personification of the qualities of the archetypal maker. *Ktr*, a root word likely relating to metal, also denotes creativity, both in the sense of craftsmanship and of natural generativity, including childbirth.[430] Hasis, from *hss*, "wise one," is a

---

[425] Bachelard, *Earth*, 136.
[426] Babatunde Lawal, *The Gèlèdé Spectacle: Art, Gender and Social Harmony in an African Culture* (Seattle: University of Washington Press, 1996), xvi.
[427] Gantz, 365.
[428] Bachelard, *Earth*, 138, n. 74.
[429] Ibid, 137.

praise-term also applied to the god Ea in the *Enuma Elish* as well as to Utnapishtim in the *Gilgamesh* epic. Hephaistos too is "praised not only for his craftsmanship but with intellectual epithets.... He is 'very wise,' 'renowned in wisdom,' 'rich in wisdom,' 'with knowing heart.'" Like Hephaistos, Kothar-wa-Hasis possesses magical powers. Some Ugaritic sources "compare Kothar with figures like Ea, the Mesopotamian god of magic."[431]

It is not possible to say where or how Hephaistos originated, but the overlapping of Kothar and Hephaistos may be quite ancient, going back to Crete, where his presence may be attested in an inscription at Knossos although this is in dispute.[432] The Ugaritic god Kothar, whose origin-place lies in the region of modern Syria, also has the "throne of his dwelling" on Crete.[433] Greek myth gives Hephaistos Cretan connections, including Zeus's gift of the bronze man, Talos, as a servant to the Cretan king, Minos. Kothar's appellation as the "wise and skillful one" has its Greek linguistic equivalent in Daidalos. Thus, observes Morris, "Daidalos would be an *interpretatio Homerica* of Kothar...perhaps the result of an epic collision with a craftsman already native to the Aegean and its poetic tradition." This tradition, probably originating in the Middle East, may well represent a "two-way street" of trade and influence driven by the diffusion of metallurgy going back to the early Bronze Age. According to Morris:

> The talents and techniques of Kothar, Hephaistos, and the semantic range of *daedala* are embraced in the Biblical concept of an early Iron Age craftsman, whose talents in wood, textiles, metalworking, and jewelry are the most highly valued. Before Daidalos evolved into a vehicle for classical and philosophical values [in fifth-century Athens], he belonged to this artistic sphere, much as the Mycenaean world shares more with the Near East than with its classical heritage.[434]

Both Hephaistos and Kothar-wa-Hasis are associated with powerful female goddesses of generativity. A group of goddesses called "the *ktrt* or 'skilled ones,' have been interpreted as both

---

[430] Morris, 81.
[431] Ibid, 85-87.
[432] Ibid, 77.
[433] Ibid, 93.
[434] Ibid, 97-98.

guardians of marriage and childbirth and as singers or musicians."[435]
It has been argued that this name-association is a function of the
collaborative relationship of the female *ktr*-deities, who preside over
"the creation of life in nature" as Kothar presides over artisanry.[436]
Loraux asserts that the two Eileithyiai, the goddesses of birth who
appear on either side of Zeus in Athenian vase paintings depicting the
birth of Athena, "compete" with Hephaistos when they do not indeed
entirely supplant him. She points out that the texts describing the
birth "generally resort to the language of generation, the part of the
woman (*tiktein*: to bring forth) balancing that of the man (*gennān*: to
beget)." However, she adds, "We should not be surprised to find that
this vocabulary enters into a competitive relationship with another
semantic sequence, which employs the artisanal language of the
metallurgical process to speak of the birth."[437] As was noted in
Chapter 2, Heidegger demonstrates the sense in which *technê* is both
a "bringing forth" and a "revealing" of things in their true nature. In
view not only of the example of Kothar-wa-Hasis and the Ugaritic
creatrix goddesses but also of the association of Hephaistos with the
Great Mother in mystery cult, it may be asserted to the contrary that
the larger mythic picture suggests an example of collaboration, not
competition, in these images.

Kothar-wa-Hasis is instructed to make precious objects for Lady
Athirat of the Sea, the "creatress of the gods:"

> The Clever One went up to the bellows,
> the tongs were in the hands of Hasis;
> he poured silver, he cast gold,
> [...............................]
> A divine throne with seat in gold,
> a divine stool covered with electrum
> divine sandals with straps
> which he has plated with gold
> a divine table filled with figures
> creatures of the foundations of the earth....[438]

[435] Ibid, 89.
[436] Ibid, 90.
[437] Loraux, 132.
[438] Morris, 80-81.

The rich design of the silver chair that Thetis sits upon when she alights at the bronze mansion of Hephaistos reflects not only a similar excellence of craftsmanship and luxury to that of the throne made for Lady Athirat, but also the similarly exalted position of she who occupies it.

Similar to the "table filled with figures" created by Kothar is the *daidalic* crown Hephaistos makes for Pandora with all the animals of Earth depicted on it. Aside from expressing the intermingling of the images of the smith gods in the Mediterranean by the late Bronze Age, such productions show the whole created material world to be the mythic territory of the smith, a theme that appears in many mythologies. An eighteenth-century royal brass Benin stool depicts the cosmos, "represented by the sun, the moon and the cross, a Benin symbol of creation" and "wild animals representing...the powers of the forest...an image indicating terrifying supernatural powers." The animals also represent the healing wisdom of the forest. "In the middle zone are symbols of civilization, and these are—not surprisingly—the products of the smith."[439]

The smith enjoys what Eliade terms a "mystic fraternity" with kings, both mythically and culturally. The Fans of West Africa made no distinction between "the chief, the medicineman and the smith, because smithcraft is so highly honored that only chiefs and their kin are allowed to ply it."[440] The smith is a civilizer who serves the ruling deities by providing the material symbols of their power. This includes both their irresistibly powerful weapons as well as many other awe-inspiring trappings expressive of the ruler-god's status, like the palace Kothar-wa-Hasis conceives and builds for Baal and the brazen mansions Hephaistos creates for the Olympian gods.[441] Hephaistos makes the scepter that passes from Zeus ultimately to Agamemnon, emblem of his inherited terrestrial power. Ogun forges the regalia of royalty, including the sword that is the symbol of royal legitimacy for the Yoruba. "The iron sword of Ogun was perhaps his most meaningful symbol, for it condensed the twin meanings of

---

[439] Sandra. T. Barnes and Paula Girshick Ben-Amos, "Ogun, the Empire Builder," in *Africa's Ogun: Old World and New*, ed. Sandra T. Barnes (Bloomington: Indiana University Press, 1997), 55.

[440] R. J. Forbes, *Studies in Ancient Technology*, Vol. 8 (Leiden: E. J. Brill, 1964), 54.

[441] Ibid, 83.

aggression and civilization. It cleared the forest and built the house," creating the conditions for war and the conditions for civilization. The establishment of royal legitimacy is symbolized by placing the "Great Sword, the Sword of Justice" in a king's hands upon coronation, followed by a visit to the shrine of Ogun.[442] By virtue of his connection with his patron god, the mortal blacksmith is honored very widely throughout West Africa. The objects made by the blacksmith "decorated the royal palace and ritually enhanced the power of the king," and "All smiths were believed to control arcane knowledge, which they used for the enhancement and protection of the nation."[443] And,

> Wherever they settled, ironworkers acquired significant ritual status. Their forges and smelters were seen as ritual shrines or sanctuaries for anyone losing a fight or fleeing turbulence. The anvil was widely used for taking oaths and as a sacrificial altar.[444]

Similarly, when making a *kris*, the knife whose powers are mythic and awe-inspiring—in short, *daidalic*—the Javanese *kris*-smith converts his smithy into a *taroeb*, a sacred space, that becomes a meeting-place for the community, and he ritually consecrates the making with the same offerings that pertain to circumcision rites and wedding ceremonies.[445]

The smith is known as the wise one, the possessor of wondrous skills and magic arts. The smith may be regarded ambivalently because of this power, but performs a function that is necessary to human life. The medieval smith says, "How does the ploughman get his plough or his ploughshare, or his goad, but by my craft? How does the fisherman obtain his hook, or the shoemaker his awl, or the tailor his needle, but by my work?"[446] Indeed, it is necessary to the power of the ruler gods, for from whom do they receive their weapons but the smith?

---

[442] Barnes and Ben-Amos, 58.

[443] Ibid, 46.

[444] Barnes and Ben-Amos, 52.

[445] Forbes, 70-71.

[446] Ibid, 59.

## *Hephaistean Fire*

Hephaistos is the god of fire. Lewis Richard Farnell notes phrases in ancient Greek literature such as the "Homeric phrase describing the cooking of meat, 'they held it over Hephaistos.'" In *Antigone*, the prophet "says of the offerings that refused to catch fire on the altar, 'from the sacrifice Hephaistos did not gleam....'" Also in *Antigone*, the fire used by the enemy to threaten the walls of Thebes is called "the Hephaistos of the pine-torch." Aristotle reports that when the fire crackles on the hearth, "Hephaistos is laughing." Farnell sees these anthropomorphic images as relics of a primitive religious perception that sees divinity in natural objects and personalizes them. Undoubtedly, observes Farnell, one of the characteristics of fire, its "weak and wavering movement," suggesting the lameness of the fire- and smith-gods (Agni, Wieland, Hephaistos).[447] Other gods, too, live in the fire: Hestia in the sacrosanct home-fire of the hearth, Hermes in the flickering and changeableness of fiery flames, Apollo in the fire of the oracular tripod cauldron, Dionysos in "fire miracles."[448] All of the Greek gods were worshipped with fire-offerings. What then is the character of Hephaistean fire?

Hephaistos was celebrated in fifth-century Athens by a ritual torch-race, the Lampadephoria, whose object was to "pass a lighted torch from hand to hand in the quickest time from the starting-place to the goal," the goal being an altar lamp which must be lit by the torch of the victorious runner. The Lampadephoria may have been a feature of three different festivals in the Athenian calendar. In at least one, the starting place was the altar of Prometheus, and the end may have been the altar of Athena Polias on the Acropolis. In the festival of the Apaturia, magnificently costumed Athenian citizens lit torches from the hearth and sang hymns to Hephaistos while sacrificing to him as the god "who taught the use of fire."[449]

Another ritual involving fire occurred on Lemnos, the island of Hephaistos (and may have persisted into the twentieth century). Philostratos tells of a ritual whereby all hearth fires were

---

[447] Farnell, 374-75.
[448] Burkert, 60.
[449] Farnell, 380.

extinguished for nine days, after which they were rekindled by fire sent by ship from Delos—in other words, rekindled by the sacred fire of Apollo.[450] Thenceforth, a "new life" would begin for the Lemnians, "especially for the craftsmen who depend on fire...."[451] The ritual was meant to expiate for the "crime" of the Lemnian women, who murdered their husbands because they had taken foreign women into their beds after Aphrodite cursed the Lemnian women with a bad odor because of their neglect of her. It is not told what the nature of this "neglect" may have been. (It is tempting to wonder whether Hephaistos is implicated.) If the fires are restored by the purifying fire of Apollo, whose fires were extinguished and why? Farnell remarks that "before the Delphic Apollo won his prominence," the natural source for the fire brought to purify the inextinguishable lamp on the altar of Athena Polias "would be considered the altar of Prometheus and Hephaistos, for both were givers of divine fire, and therefore their altar would be considered the sacred fountainhead of it."[452]

The qualities of fire are many, and Hephaistos represents not all fire, but specific kinds and qualities of fire. Hephaistos, first and foremost, represents the powerful fire of the forge, controlled through the blacksmith's magic and *technê*. He also represents telluric fire—the thunderous noise and explosiveness of molten volcanic magma is recalled in the fire of the forge, in which the hardest and coldest of substances are startlingly transformed into dangerously red-hot, malleable materials which only the blacksmith is capable of bending to his will. A cult center to Hephaistos existed on Lipari, amid the volcanic islands, the ancient "Isles of Hephaistos," off the northeast corner of Sicily, not far from Etna, one of the most powerful and violent volcanoes on the planet. Though Mount Moschylos on Lemnos, the center of the oldest known cult to Hephaistos, has not been active for many centuries its smoke is thought to recall memories of active eruption, and certainly signified the smoke of flames rising from the god's smithy.

---

[450] Ibid, 384.
[451] Burkert, 61.
[452] Farnell, 385.

Bachelard points out, however, that "...volcanoes are rather rare to have given rise to so many reveries of subterranean forges." And so perhaps," he suggests, "we would be better off if, instead of listening to mythologists who know, we listened to mythologists who reimagine." Thinking that Lemnos is the location of Vulcan's forge because of its volcano is "mythology after the fact." Indeed, his cult on Lemnos may have had more to do with sea than smoking mountain. "Spontaneous mythology, like that of Homer, locates Vulcan's forge rather in the heavens."[453] In fact, tradition told that the sky was made of bronze, and Proclus affirms that "whoever made heaven was a smith." Eusebius compares Hephaistos' ovoid *pileus*, which is blue, to the vault of heaven.[454] In the *Kalevala*, it is the divine blacksmith Illmarinen who "forged the heavens / And the arch of air who welded."[455]

So it seems that fire is the answer to only part of the imaginal riddle of Hephaistos, whose connections to water, earth, and air are also profound. An example from another mythology serves to assist in a mythical "reimagining" of the nature of Hephaistean fire. The ritual praise and propitiation of the West African blacksmith god Ogun clearly involves the image of iron as both red-hot and malleable and cold and hard. Therefore, like the iron he masters through stages of transformation, Ogun is praised as a god who is capable of containing the irreducibly opposite qualities of fiery and cool. The divine blacksmith archetype is two-sided, paradoxical, a container of opposites, just as fire as image contains both destructive and helpful qualities. Although Hephaistos does not originally work in iron, but rather in the noble metals—silver, gold, bronze—his myth contains the implacable as well as the paradoxical qualities of iron, qualities that are shared by the divine blacksmith lineage. Hephaistos, the gently persuasive peacemaker of Olympus, is equally capable of casting his *phlegma kakon*—"deadly flames—on the river god Xanthus as he is of forging the most delicate of *daedala* to adorn the necks and arms of goddesses. Ogun, whose nature is said to be iron as Hephaistos' is said to be fire, may be experienced as violent

---

[453] Bachelard, *Earth*, 125.
[454] Delcourt, 62-63.
[455] Bachelard, *Earth*, 138.

or beneficent toward humankind. In Ogun's varied rituals, his devotees do not seek to reconcile the seeming opposites of the god's destructive or humanitarian qualities. Instead, they say, "Ogun has many faces."[456]

Ogun, who is maker of the sword, is also the patron of the body-scarification artist (*olóòlà*). The ideal *olóòlà* has "a cool, patient character. He must not be excitable but friendly, attractive and approachable. He must not drink liquor or palm wine, because if he drinks a lot he won't be able to do the marks well." The irreversibility of the *olóòlà*'s art is reflected in the praise with which he invokes his god: "The scar, the road, and Ògún are alike. None can be changed."[457] The smith is not a god of wine. Neither Ogun nor Hephaistos drinks—when each does become drunk, the result is mythic. Bachelard speaks of the phantasmagorical "Hoffmannian fire" of "spontaneous combustion," fueled by the flame of alcohol. "Bacchus is a beneficent god; by causing our reason to wander he prevents the anchylosis of logic and prepares the way for rational inventiveness."[458] "Anchylosis" means "stiffening of the joints," from the Greek root *ankulos*, ("crooked," "bent").[459] This is also the root of one of Hephaistos' epithets, "crook-foot." An interesting question for further study would be: how is Dionysian fire expressed in the drunkenness of Hephaistos?

Anger is often compared metaphorically to fire. Yet, the anger of Hephaistos seems above all, like his forge-fire, subject to his will. When Helios tells Hephaistos of his cuckholding, Homer tells us that "His heart [is] consumed with anguish," but the story reveals how the god's anger contains cool calculation. His rage is quickly cooled once his plainly stated demands are met for the return of Aphrodite's bride price and a guarantee that the damages owed by Ares will be paid. He expresses rage toward "bitch-eyed" Hera in the most matter-of-fact way to Thetis, then turns to the work at hand of making the Shield.

[456] Sandra. T. Barnes, "The Many Faces of Ogun: Introduction to the First Edition." In *Africa's Ogun: Old World and New*, ed. Sandra T. Barnes (Bloomington: Indiana University Press, 1997), 2.

[457] Drewel, 256-257.

[458] Gaston Bachelard, *The Psychoanalysis of Fire*, trans. Alan C. M. Ross (Boston: Beacon, 1964), 87.

[459] TheFreeDictionary.com, "ankulos," Farlex, 2004, accessed November 20, 2004, http://www.thefreedictionary.com.

Homer gives no sign that the artisan's craft is fueled by the emotion of anger. Instead, the image is of his exquisite control of the bellows, which can blow with force or the faintest breath at a moment's notice. All attention is turned to the work. Rather than the blacksmith, in the poetic language of the *Kalevala*, it is *steel* that is "seized by 'anger' and iron by 'fury' in the cooling trough." Bachelard speaks of "the idea of imprisoning fire in iron with water, the idea of caging the savage beast that is fire in a prison of steel...."[460]

Bachelard comments on the "reverie of dark fire, so common in the writings of the alchemists." He notes its association with the image of the glowing charcoal, which ranges in color "from darker to brighter red" and is brought back to life—made redder—by the breath.[461] Eliade notes that, "The 'mastery of fire,' common both to magician, shaman and smith, was, in Christian folklore, looked upon as the work of the devil." It will be remembered that Milton chooses to place Mulciber, another Roman name for Hephaistos, in Hell as Lucifer's architect. Alchemy and sacrifice in connection with forge-fires each deserve full-length studies. With regard to the archetypal blacksmith's mastery of fire, Eliade cites the example of Odin-Wotan, who is "master of the *wut*, the *furor religious*" as an example of,

> this intimacy, this sympathy with fire, which unites such differing magico-religious experiences and identifies such disparate vocations as that of the shaman, the smith, the warrior and the mystic.

Eliade cites themes in European folklore of "Jesus Christ (or St. Peter, St. Nicholas or St. Eloi)," who "heals the sick and rejuvenates the old by putting them in a heated oven or forging them on an anvil." It is in this connection too that the Gnostic "baptism by fire" may be understood. Jesus is thus the "'master of fire' par excellence."[462] These and other tales emphasize the blacksmith and his archetypal fire as agents of magic, transmutation and profound sacred power.

---

[460] Bachelard, *Earth*, 117.
[461] Ibid, 119.
[462] Eliade, *Forge*, 106-07.

CHAPTER 5

# The Return of Hephaistos: Re-Mything Art and Technology

HEPHAISTOS today is seldom invoked by name. However, the archetype he represents is to be seen everywhere in contemporary life. This chapter will identify and examine readily recognizable instances of the Hephaistean mythos animating ideas that appear in a variety of cultural forms.

Two specific themes emerge from such an examination. One I will call the "Wounded Artist," whose problematic mother-relationship issues in an impairment of creativity. The other is that of "Monstrous Technology," expressed in artistic and other productions that image unconscious fears of divine and human creative capacities, fears that can be shown to be rooted in very ancient origins.

That these themes are so prevalent does not indicate a misreading of the myth of Hephaistos: they *are* the myth as we live it today. This fact answers the question that Jung raised: "What myth is living me?" At least, so far as the Hephaistean archetype is concerned. It also raises a further question: what are the implications of recovering the fullness of the image of the divine blacksmith-creator god: his magic, his ability to contain and represent the tension of opposing energies, and his fruitful woundedness, in both its dark and shining aspects? Archetypal psychology has much to offer on the subject of how myth operates in and on the psyche, and how we may live with and through mythic images in a way that is both faithful to the image and creative of new psychic and material forms. In this connection, I will discuss James Hillman's ideas concerning what he terms "soul-making" and Mary Watkins' concept of imaginal dialogues.

Finally, I will present examples of Hephaistean re-mything. How can the artist function effectively in society, speaking his or her *mythos* in a powerful and constructive way, as powerful as the civilizing legacy of the divine blacksmith-artisan? How may technology use the most awesome and radical discoveries of science, knowledge that offers the power to manipulate the very stuff of life on this planet? How might the artist and technologist, both members of the lineage of the divine smith, claim their creative powers in their daylight as well as their shadow aspects, honoring both the wound and its magic, and the unique instrumentality it confers; an instrumentality upon which the other gods must rely for material creation to be continually renewed and enriched?

## The Wounded Artist

Depth psychologists in the twentieth century evidenced a central interest in myth as expressive of features of the human psyche observable in the therapeutic-analytic setting. For depth psychologists like Murray Stein, Hephaistos is specifically and uniquely representative of the shadow aspect of the artist in culture—introverted and marginalized, socially and sexually suspect, marked, different. He is the holder par excellence of the "wound" that separates a man from his creative instinct.

In 1910, Freud wrote a study titled "Leonardo da Vinci and a Memory of his Childhood." Though Freud never mentioned Hephaistos, he chose as the subject of his study the single most paradigmatically Daidalic, and thus Hephaistean, artist since classical times. Freud's conclusions have been enormously influential on the psychological view of the artist that prevails in Western culture. According to Giorgio Vasari, Leonardo "in the last hour of his life...reproached himself with having offended God and man by his failure to do his duty in his art."[463] Freud's central argument is that

Leonardo's artistic creativity was inhibited and that this condition originated in a mother problem. The weight of the analysis depends upon the supposition that Leonardo—who was illegitimate and whose presence in his father's household is not documented until he was aged 5, well past the critical period of infancy—experienced a precocious mutual bliss with his mother during his suckling period. For Freud, the energetic scientific researches which are recorded in Leonardo's extensive notebooks together with his inability to bring all but his earliest works to a state of completion are an expression of the pressures of his unconscious and unrequited sexual instinct, which he was unable to fulfill through the sexual love of either men or women. Says Freud, "We must be content to emphasize the fact—which it is hardly any longer possible to doubt—that what an artist creates provides at the same time an outlet for his sexual desire."[464] In this statement, the legendary artist Leonardo, and with him the image of the artist generally, attains to a newly *revalorized* myth, to use Bachelard's term.[465]

Bachelard writes that "even those who love legends are capable of lessening their resonance."[466] Such has been the case with Hephaistos, the divine artist and craftsman. Jungian analyst Irene Gad recalls that, in spite of Hephaistos' masculine image in ancient art, depicted with "vigorous arms, a powerful neck, a broad chest with compact muscles, a mighty, forceful figure," he is nevertheless most often recalled as the "butt of bedroom farce," and that "derogatory comments about his lack of grace and clumsiness seem to prevail."[467] "What anxieties and insecurities in us," she wonders, "have made him an object of derision?"[468]

## The Mother-Wound

"What man among us has not been wounded in his creativity?" asks Murray Stein in a series of taped lectures on "The Hephaistos

---

[463] Georgio Vasari, *Lives of the Artists*, qtd. in Freud, *Leonardo*, 14.
[464] Freud, *Leonardo*, 82.
[465] Bachelard, *Earth*, 129.
[466] Ibid, 127.
[467] Irene Gad, "Hephaestus: Model of New-Age Masculinity," *Quadrant. Journal of the C. G. Jung Foundation for Analytical Psychology*, 2.3 (Fall, 1986), 32-33.
[468] Gad, 33.

Problem: An Exploration of the Wounded Creative Instincts." Where Freud implicates his mother's damaging love in the regrettable weakness of Leonardo da Vinci, Stein implicates the "patriarchy" in the inability of men as well as women to express their innate creative instinct. Stein observes of the men he has encountered in the therapeutic setting that he cannot think of one that does not have "strong Hephaistean features, that wasn't wounded in his area of creativity at some point in his early life, in a very paralyzing and injuring way, albeit seemingly very small. And...it has to do with being shamed...."[469]

The source of this shaming for the mythical Hephaistos is his violent rejection by his mother Hera when he is born lame, says Stein. Later he is ejected again from Olympus, this time by the father, Zeus. "In both stories," Stein says, "Hephaistos is victimized by the bitter and conflicted relations between Zeus and Hera, the divine couple of classical Greece and the representatives of archetypal father and mother." More than that, says Stein, Hephaistos is the product of that conflict. Misshapen, he "therefore represents a symptom or problem in Greek culture."[470]

This problem lies in the patriarchy. Stein regards Hera as representative of the frustrated mother who, stripped of her primordial power by the patriarchy, "conceives in anger and gives birth with vengeance."[471] Hera's "project" in conceiving Hephaistos absent the father is "narcissistic." According to Stein, "We all have a Hera. Hera represents the collective consensus...the dominant attitude of society—she's the queen of heaven." Hera is "defensive" and "threatened by the creativity of others." As a female subject to the patriarchy,

> she's been damaged and humiliated herself—her creative role has been usurped, she's no longer the young, strong creative force that she once was. Now she's established, she's the wife of a dominant paternal power of a patriarchal society, and she is that structure in all of us that is anxious about change, and creativity is all about change. She's that part of us that feels the threat of loss if something new

---

[469] Murray Stein, "The Hephaistos Problem," rec. July 10, 1993, C. G. Jung Institute of Chicago, audiocassette #519: tape 2 side A.
[470] Stein, "Hephaistos Problem," tape 1 side B.
[471] Ibid, tape 1 side A.

comes into being.... This is a dominant anxiety, a crippling anxiety that thwarts the possibility of a new creative expression that may appear in ourselves and casts it aside. She's the anxious mother in a man's psyche that says "color in the lines"...and yet she cannot prevent the strong workings of the creative instinct from below.[472]

Hephaistos has been "cast out of his parents' household and out of the structures of dominance...and forced to live underground in caves and out of the way places."[473] Stein points out that the Greek word for the hidden place where Thetis and Eurynome take him is *mukos*, which means "the innermost place," "the secret place," but also "the women's apartments in the house." It is here that Hephaistos learns the arts of metallurgy that belong to the Great Mother.[474] In this feminized landscape, however, Hephaistos' art cannot mature. He spends nine years of "incubation" making what Stein dismisses as "baubles" for his foster-mothers. (Homer instead uses the fateful term *daedala* to designate these objects.) Then, the angry, resentful son fabricates the golden throne that entraps his mother Hera. Returned to Olympus only through the intoxicating effects of the wine of Dionysos (for, "Wine douses fire," remarks Stein), Hephaistos releases Hera—i.e., releases his rage toward her—and they are reconciled. Stein tells us that this forgiveness and reconciliation must occur for the son to begin to free himself from the creative wound inflicted by his mother's narcissistic rage. If this release were not to occur, says Stein,

> the hardening in this attitude of irreconcilable bitterness against his mother would spell absolute disaster for both, for it would leave the mother suspended and suffering in mid-air and the son cut off from the creative energies that flow into him through his contact with the Mother. Caught in his own trap, Hephaistos would wither in the self-destructive heat of resentment.[475]

Stein invites us to see that by the time of the banquet in Book 1 of the *Iliad*, Hephaistos has been sufficiently reconciled to Hera that he behaves as a fond son. He is careful in de-fusing the Father Zeus's rage, mindful of punishments that both he and his mother have

[472] Ibid, tape 1 side B.
[473] Stein, "Hephaistos Problem," tape 1 side B.
[474] Stein, "Pattern of Introversion," 42.
[475] Ibid, 46.

received from the violent father. He has not forgotten the hurt received from her, as his speech to Thetis in Book 18 attests; but he is in control of his choices, as he is, at last, of his art, as evidenced by the *magnum opus* of the Shield of Achilles. Finally, returned to Olympus and reconciled to his mother, he has been able to realize the potential of his gifts, only then becoming the renowned architect and artist of Olympus.[476]

Myths are confusing—witness the myriad schemas for their interpretation. However Stein's portrait of the Hephaistos image shows that the literalization of myth in the "scientific" tradition developed within psychology may be more confusing still when one attempts to follow its logic, for the interpretations of both Stein and Freud before him are based on logical premises. They attempt to apply to a mythic figure principles of human psychology—in Stein's case to the Greek god Hephaistos, in Freud's to the near-mythic artist Leonardo whose scanty biography provides few points of evidence on which to base an analytical case-study—*rather than allowing the image to speak its mythos*. The image of the artist is not allowed to speak for itself.

Stein portrays Hephaistos as "the quintessential fringe-person on Olympus." He paints a detailed portrait of the attitude of the god who returns after being cast from Olympus:

> Included at the edge, he looks uneasily in, into the wheels-within-wheels that make up the Olympian social structure.... Trying somehow to stay in touch with the center, maybe to be ready for the worst or to know what's coming next, he knows all the while that it's impossible really to belong there—there, where they tolerate the fringe-people as long as the work gets done, but where they can never act and feel quite easy and neighborly with them.[477]

Further, "the only Olympian god who works," Hephaistos bears "something of the mark of an inferior child who has to take up a trade." He "...stands on the fringes of the power circles that govern the Olympian world, a servant-artisan figure who builds the palaces of the Gods 'with skillful hands' and sometimes plays the court buffoon to the great amusement of his fellow Olympians."[478]

---

[476] Stein, "Hephaistos Problem," tape 1 side B.
[477] Stein, "Pattern of Introversion," 36.
[478] Ibid, 35.

The portrait that is being painted here and titled "The Hephaistos Problem" is not so much a portrait of the god as a portrait of how at least one depth psychologist reflects back the portrait of the "wounded" artist as it already resides within twentieth-century society: above all, the marginalized artist is dispossessed of legitimate power. At best he is a laughable buffoon. And the root cause of this malaise, according to the studies of Freud and Stein, is the identification with the feminine on the part of their respective subjects.

Stein's statement that Hephaistos comes into his full creative power only within the framework of the patriarchal structure that he faults in the first place is an internal contradiction within his argument, and must therefore be acknowledged as invalid. Here is Stein's portrait of what he terms the "Hephaistean" man, based on observations in therapeutic sessions that Stein links to the image of the god:

> He will presumably find himself rather an outcast from a conventional world that requires ready adaptation to patriarchal and masculine dominants; he will be moody and given to swinging between inflation and depression; he will appear both to himself and others, especially to the analyst, rather unheroic and uninterested in overcoming his close attachment to the world of women and mothers; indeed he will cling to feminine circles and company, fascinated by the mysteries of creativity and often lost in a world of inner images and fantasy, bound hand, foot and soul to the excitement and anguish of tending the "underground forges." He will seem to be quite anima-possessed, smoldering and crippled.[479]

In Stein's statement of the "Hephaistos problem," to the degree that the artist is shown to be feminized, he will be seen to be ineffectual in his most important function, that of making things that matter, i.e., things that are recognized to be of value to the patriarchy. The weight of implication for the responsibility for the wound to the creative masculine ego falls not on the patriarchy but on the mother, the Mother, and the anima.

---

[479] Stein, "Pattern of Introversion," 42.

## Negative Anima

Erich Neumann's privileging of the "solar" over the "chthonic" phallic is indicative of a revaluing, in Bachelard's sense of the term, of the profound mythic connection between the Great Mother and the chthonic phallic powers. These powers have a deep and ancient Hephaistean connection, as has been seen in the persistence of the myths and legends of the magically creative, ancient, chthonic phallic divinities such as the Kabeiroi, Daktyloi, and Telchines. However, the bias toward the characteristics of the paternalistic, solar masculine is evident in Freud's diagnosis of the imputed homosexuality of Leonardo as being both negative in its impact on his life and feminine in its origins and effects.

Stein's Hephaistean man "smoulders" with creative fire he is unable to express. As Stein rightly points out (following Farnell), Hephaistos the god *is* fire. However, I suggest that Stein's version of Hephaistos' fire, which can shine but dimly until it gains the approval of Zeus, can be identified with Neumann's "chthonic" phallic, while Zeus's patriarchal Olympus is an example of the "solar" phallic. The chthonic phallic has long been lurking in the shadow of depth psychology, and at the root of its disparagement is a de-valuing of the pull to the feminine, which was demonized as incestuous and as weakening the *membrum virile* of masculine ego. As Stein says, "The loves of Hephaistos, which begin with Aphrodite, tend to conclude in disappointment. Behind these disasters lies the incestuous, Mother-directed motion of his libido."[480]

It must be noted here that "creativity" as both Freud and Stein use it is tacitly defined by the action-addicted patriarchal culture, which fails to sufficiently value the stages of reflection and "incubation" in the creative process. Emma Jung, cited by Stein, makes the following statement on the function of the anima:

> The transmission of the unconscious contents in the sense of making them visible is the special role of the anima. It helps the man to perceive these otherwise obscure things. A necessary condition for this is a sort of dimming of consciousness; that is, the establishment of a more feminine consciousness, less sharp and clear than man's

---

[480] Ibid, 47.

but one which is thus able to perceive in a wider field things that are still shadowy.[481]

What is being described is a positive characteristic of creative process which does not always lead to the production of tangible works, but without which the work of art cannot be produced. Yet, Emma Jung's description of the "dimming" of "feminine consciousness" in reverie recalls Freud's negative attribution to Leonardo of a "dim notion of a perfection" that he believed inhibited Leonardo's ability to complete the Last Supper fresco in Milan. Freud contrasts this to Leonardo's supposed earlier "masculine" period of decisiveness and activity before he was overcome by his infantile history of mother-bliss.[482]

In Jungian psychology, it is the anima that gives voice to the unconscious in a man. Yet, by his own account, Jung rejected the voice of his own anima during his period of psychic distress and personal transformation following catastrophic dreams and waking visions in the years following his break with Freud (1913-18). In *Memories Dreams Reflections*, he is recorded by Aniela Jaffé as saying that this voice told him that what he was doing in exploring his own psychic processes was "art." Jung's response to this is to accuse the interior female voice of his anima of deep cunning:

> If I had taken these fantasies of the unconscious as art, they would have carried no more conviction than visual perceptions…. I would have felt no moral obligation toward them. The anima might easily have seduced me into believing that I was a misunderstood artist, and that my so-called artistic nature gave me the right to neglect reality.[483]

Jung discredits the seductive *logoi* of this feminine voice and negates the connection of the artist to "reality." This implication has distinctly negative reverberations in relation to the image of the artist. Yet Jung's healing process began and continued in daily artmaking, beginning with his resumption of a childhood activity of building play towns with small stones, an activity he undertook in response to what he took to be an invitation from his unconscious. Jung however claims that while his healing process was occurring, "I gave up this

---

[481] Ibid, 43.
[482] Freud, *Leonardo*, 16.
[483] Jung, *Memories*, 187.

estheticizing tendency in good time, in favor of a rigorous process of [scientific] understanding."[484] Yet, it must be noted that Jung's illuminated journals, the Black and Red Books, with their archaic calligraphy and beautiful, perhaps obsessively complex illustrations comprise a startling 1,330 pages.[485] What this underscores is the value the artist intuitively and often consciously assigns to the *process*, not the product, of artmaking as an activity that seeks and itself propels the development of consciousness. Late in life Jung may have dismissed the enormous energy he invested in the Black and Red Books in his younger years, but he did not in fact give up art making as process: "Any time in my later life when I came up against a blank wall, I painted a picture or hewed stone."[486] Still, the image of the anima is made to sit uneasily in relation to the moral exigencies of "reality."

## Negative Animus

As Jung indicts the male's creative anima, Stein implicates the female's creative animus. Stein draws a strong distinction between masculine creativity and what he terms "simple, natural feminine creativity." He asserts that Hephaistos connects a creative female to "her deepest feminine-maternal impulses, yet wants something other than simple maternity." The Hephaistean animus tends toward the creation of artificial products and thus Hephaistos may be "a monstrous offense to feminine naturalism, a sick-making disharmony in the tones that vibrate between feminine ego-consciousness and the Great Mother."[487] Reminding his hearers that Hephaistos is *contra naturam* – "His feet go backwards"—Stein warns of the threat the Hephaistean animus represents to the creative feminine psyche, which he asserts prefers "natural" creativity to artificial creation. The inescapable conclusion here is that the only effective use of women's creativity is limited to what they can produce and nurture out of their bodies. Further, "artificial" creation seems to be denied to females as "unnatural." What would this mean, for example, to the female

---

[484] Jung, *Memories*, 188.
[485] Robert C. Smith, *The Wounded Jung* (Evanston: Northwestern University Press, 1996), 76.
[486] Jung, *Memories*, 175.
[487] Stein, "Hephaistos Problem," tape 2 side B.

sculptor or indeed any woman who might choose to work creatively in the fields of biotechnology or artificial intelligence?

My purpose in critiquing Freud, Stein, and Jung is to point out the persistence, in the writings I have cited, of a particular attitude toward art and artists, certainly discernible in the culture-at-large at least since the Romantic era, which Stein has usefully identified with the image of Hephaistos. This attitude negatively identifies the male artist as mother-wounded and the female artist as unnatural, drawn away *from her grounding with the Mother. Thus pathologized, both become culturally suspect.*

## Deformed Perspectives

James Hillman redefines the term *pathologizing*,

> to mean the psyche's autonomous ability to create illness, morbidity, disorder, abnormality, and suffering in any aspect of its behavior and to experience and imagine life through this deformed and afflicted perspective.[488]

What, Hillman asks, might the soul be *imagining* by means of pathology? The question is not how to reconcile the mythic symptom into the waking world—in the case of Freud's Leonardo and Stein's Hephaistos with the affirming values of the patriarchy—but how to dream into history. By "history" I mean what Hillman speaks of as the "digestive operation" of "the history-making of the psyche, sifted and weighed in the disciplined reflection of loving, of ritual, of dialectics, of an art—or of a psychological analysis with its therapeutic plot." In the case of the Hephaistos image, the question becomes: what is the *mythos* of the wounded god-image saying—and how can this be digested into the individual history of the artist, or indeed the creative maker within the soul of anyone who is being worked on by this archetypal energy? This question is not to be posed in a literalizing way, but rather digested by transforming psychic events "from case material to subtle matter."[489]

Jung's attitude of rejection toward his anima denies that the artist-image is connected to reality and strongly implies its disconnection from the everyday world of ethical values. By contrast, Hillman

---

[488] Hillman, *Re-Visioning*, 57, italics mine.
[489] James Hillman, *Healing Fiction* (Woodstock: Spring, 1983), 27.

claims that for archetypal psychology, as differentiated from analytic psychology, soul-making is located "in the world." "More specifically," states Hillman,

> the act of soul making is imagining since images are the psyche, its stuff, and its perspective. Crafting images...is thus an equivalent of soul-making. This crafting can take place in the concrete modes of the artisan, a work of the hands, and with the morality of the hands.[490]

This "morality" emerges from what Kenton S. Hyatt describes, following Martin Buber, as an I-Thou interaction between the artisan and his or her medium—stone, metal, paint, camera, and photographic chemicals and paper, words—which the artist may regard as speaking its *mythos* to him or her. Hyatt notes that artists may be said to be "struggling with a medium as if it had its own volition, at times resisting, at times responding." Then, as the artist achieves a growing harmony in dialogue, and thus relationship, with the medium, the artist's work may begin to "sing."[491] Peter London writes of the "creative encounter," urging the contemporary artist to "consider the special relationship between artists and their tools," and that, "In prepared hands they yield the latent images that lie embedded in the substance to be worked." In hands "unprepared for the task, the tool uncovers nothing. Worse, it may turn on its handler, painfully reminding that person of her or his transgressions."[492] Blacksmiths traditionally regarded and revered their tools as sacred, carriers of numinous power, and prepared themselves for their work as for a sacred undertaking. The morality of the hands arises in an interaction with the stuff of the world, engaged on its own terms as a partner in the dialogue that is *making*.

## Monstrous Technology

Stein calls the artificial creativity of Hephaistos a "monstrous" affront when set beside the natural issue of biological quickening and

---

[490] James Hillman, *Archetypal Psychology: A Brief Account* (New York: Harper, 1978), 36.

[491] Kenton S. Hyatt, "Creativity Through Intrapersonal Dialog," *Journal of Creative Behavior*, 26.1 (1992), 68.

[492] Peter London, *No More Secondhand Art: Awakening the Artist Within* (Boston: Shambhala, 1989), 174.

birth. The appellation "monstrous" carries a mythic truth when seen reflected back from another mirror, that of the question of the powers of humans and the powers of gods. Artificial creation, whether it be mechanical or biological, arouses awe, fascination, and fear. In it, the power of the gods is discerned. In it, too, the powers of humans come into question.

## Mary Shelley's *Frankenstein*

In 1818, Mary Shelley published *Frankenstein; or, the Modern Prometheus*, which is viewed retrospectively as the first work of science fiction, even though the term "scientist" did not come into use until 1834.[493] Her reference to the myth of Prometheus recalls the intervention of the Titan in human creation and his dangerous gift of fire—and the punishment he suffered for daring to steal what belonged to the gods. As has been shown, Prometheus and Hephaistos are mythically inter-identified. And, it is Hephaistos who is the *maker* of problematic creatures.

Shelley's protagonist is Victor Frankenstein, whose success in bringing to life a man from pieces cobbled together and animated by means of the knowledge of profound creative forces which his tireless study has revealed to him frightens and disgusts him, and he abandons his creation. The creature finds his way to human habitation but is rebuffed as fearful, horrible, and monstrous. Isolated from companionship, he suffers the pain of human desolation and loneliness. He educates himself in human language, but for naught, as he will never be accepted but remain an outsider, the victim of a rogue creation.

Frankenstein's creation is intelligent, planful, and utterly dependent upon his creator for the fulfillment of his one desire. He eventually tracks down his creator and demands a mate. He promises to retreat forever from humankind, to whom his existence is so monstrous. Nevertheless he wishes to enjoy the "bliss" to which he, created with human understanding and passions, is entitled. "Oh! My creator, make me happy; let me feel gratitude towards you for one benefit! Let me see that I excite the sympathy of some existing thing; do not deny me my request!"[494] Frankenstein's resistance is

---

[493] Baldick, 64.

countered by the creature's reasoning and his instinct of dread is momentarily overcome by the creature's emotional appeal. But he ultimately cannot force himself to revisit the horror of his monstrous issue and cannot bring himself to duplicate it. The created man becomes murderous, killing one by one all the people Frankenstein loves: his friend, his brother, his father. Frankenstein tries to escape, but when finally his beloved is murdered on his wedding night, he becomes the implacable enemy of the monster who seems to be endowed with superhuman capabilities with which to taunt him, and they ultimately destroy each other.

Frankenstein's monster is in a sense a Pandora, an evil gift whose creation is meant to mock and punish men, diminishing their enjoyment of the gifts that would make them feel themselves to be gods. Victor Frankenstein has dared to transgress the boundary between divine and human creativity. He rails against the punishment the pitiless monster inflicts upon him. He recognizes the consequences of his action but ultimately feels no more remorse for the sorrows of his creation than do the Greek gods for those of humanity. Shelley's story says that sorrow awaits the man who takes upon himself the powers of the gods; but the gods are absent and Victor Frankenstein is left to reject or accept the fate he has created for himself. Frankenstein, having created another being without resort to sexuality is denied the privilege of sexual love and the opportunity to reproduce normally.

## E. T. A. Hoffmann's *Tales*

The tales of E. T. A. Hoffmann (1776–1822) deal with the ambiguity of the boundaries between science and sorcery, Apollonian rationality and Dionysian madness. Hoffmann's characters, however, forerunners of the "mad scientist" archetype so central to science fiction after Hoffmann, are Hephaistean. They are makers who bear the forge-mark of physical eccentricity and otherness. They are strangely proportioned. Coppelius, the sinister alchemist whom young Nathaniel confuses with the fairy-tale Sandman who tears out the eyes of children who won't go to bed, is huge, "with a head

---

[494] Mary Shelley, *Frankenstein, or the Modern Prometheus*, ed. Johanna M. Smith (Boston: St. Martin's, 1992), 125.

disproportionately big, a face the color of yellow ochre, a pair of bushy grey eyebrows, from beneath which a pair of green cat's eyes sparkled with the most penetrating luster."[495] Coppelius' *doppelgänger*, Giuseppe Coppola, the spectacle-maker, advances on Nathaniel, "his wide mouth distorted into a hideous laugh, and his little eyes darting fire from beneath their long grey lashes."[496] Hoffmann describes another of his characters, Councillor Krespel, in a manner suggestive of uncanny powers displayed through unnatural psychic attributes that express themselves in glimpses of physical difference:

> There are men...from whom nature or some peculiar destiny has removed the cover beneath which we hide our own madness. They are like thin-skinned insects whose visible play of muscles seem to make them deformed, though in fact, everything soon turns to its normal shape again.[497]

Or, more precisely, these attributes are not unnatural, but rather an externalized expression of the potential for madness that is hidden within the frail human psyche.

Hoffmann's story "The Sandman" led Freud to reflections on the power of the uncanny in fiction, musing on the German word *unheimlich* ("uncanny"). He explains that the uncanny in fiction creates its effects by throwing light on that which is meant to be hidden, uncovering it to view. Yet, the horror of the uncanny is dependent not on the simple fact of something being hidden, but of its being doubly-hidden. "The German word '*unheimlich*,'" Freud observes, "is obviously the opposite of '*heimlich*' ['homely'], '*heimisch*' ['native']—the opposite of what is frightening precisely because it is *not* known and familiar."[498] However, "From the idea of homelike, 'belonging to the house', the further idea is developed of something withdrawn from the eyes of strangers, something

---

[495] E. T. A. Hoffmann, "The Sandman," trans. John Oxenford, par. 12, *19th Century German Stories*, ed. by Robert Godwin-Jones, 1994, Foreign Lang. Dept., Virginia Commonwealth University, accessed October 13, 2004, http://www.fln.vcu.edu/hoffmann/sand_e.html.

[496] Ibid, par. 77.

[497] E. T. A. Hoffmann, "Councilor Krespel," *E.T.A. Hoffmann Tales*, ed. Victor Lange, The German Library: Vol. 26 (New York: Continuum, 1982), 92.

[498] Sigmund Freud, "The 'Uncanny.'" in *Writings on Art and Literature*, eds. Werner Hamacher & David E. Wellbery (Stanford: Stanford University Press, 1997), 195.

concealed, secret...."[499] "Thus *heimlich* is a word the meaning of which develops in the direction of ambivalence, until it finally coincides with its opposite, *unheimlich*."[500] It is this ambivalence, Freud demonstrates, which produces the *frisson* we experience in tales of the uncanny. The horror hinted at in fiction of the uncanny is the horror felt in the wake of the transgression of *heimlich* illusions. It is the same horror as in the viewer's recoil from Oedipus' revelation of his crime, which strikes at the heart of both home and *polis*, and which leads Oedipus to commit upon himself the punishment and sacrifice of self-blinding. Freud goes on to discuss the example of E. T. A. Hoffmann's story, which provides the figure of the automaton Olympia, central to the first act of Offenbach's opera *Les Contes d'Hoffmann*. It is not, he suggests, the appearance of the doll itself which is the origin of the sense of the uncanny produced in the opera, but something hidden underneath it: the allusion to the recurring theme in Hoffmann's story of the Sandman, who *tears out children's eyes* as they lay resisting sleep (or are perhaps asleep and dreaming?) in their *heimlich* beds. Freud notes that, "A study of dreams, phantasies and myths has taught us that anxiety about one's eyes, the fear of going blind, is often enough a substitute for the dread of being castrated."[501] Indeed, Nathaniel's growing madness and eventual suicide caused by his bewitchment and love for an automaton (a barren piece of wood) in a story rife with eyes—red, bloody, casting baleful and maddening rays, torn out—separates him from his fiancée Clara and his opportunity to marry and raise a family of his own. The last line of the story affirms clear-eyed Clara's survival and subsequent happy married life, hand-in-hand with her husband, her "two lively boys" playing beside her.

It will be recalled that the Telchines "harm all things with their gaze."[502]

---

[499] Grimm's Dictionary, 1877, qtd. in Sigmund Freud, "The 'Uncanny'" in *Writings on Art and Literature*, eds. Werner Hamacher & David E. Wellbery (Stanford: Stanford University Press, 1997), 200.

[500] Freud, "Uncanny," 201.

[501] Ibid, 206.

[502] Gantz, 149.

Harrison cites a verse of Callimachus on the single-eyed Cyclops that anticipates Hoffmann's themes of childhood fears wrapped in images of the home:

> Even the little goddesses are in a dreadful fright;
> If one of them will not be good, up in Olympus's height,
> Her mother calls a Cyclops, and there is sore disgrace,
> And Hermes goes and gets a coal, and blacks his dreadful face,
> And down the chimney comes. She runs straight to her mother's lap,
> And shuts her eyes tight in her hands for fear of dire mishap.[503]

This verse also reminds one of the frequency of tearful encounters between very young children and department store Santas with their bushy eyebrows, scary laugh, and *unheimlich* over-familiarity as children are familiarly dandled on the lap of a giant stranger. The association of Santa with chimney, and the present of coal to disobedient children is very ancient indeed. (And, if the European notion of St. Nick is anticipated in Callimachus' verse on the Cyclops, it should not be surprising that he is accompanied by dwarfish elves as a reminder of his chthonic origins.)

## Blade Runner

*Blade Runner* (cited by a 2004 survey of scientists conducted by the *Guardian* newspaper as the most popular science fiction movie ever made) follows the thematic path forged by Shelley and Hoffmann a century and a half previous of rogue creation connected with murder, madness, and interruption in the happy flow of normal human procreation.

Beginning in its opening visual, Ridley Scott's 1982 film presents for extended meditation the image of Los Angeles. The landscape is unmistakably Hephaistean, transformed through some cataclysmic alchemy from the artificially green and sun-drenched twentieth-century paradise to its literal shadow: infernally dark, cold, eternally raining, flushed with brightly burning plumes of flame, a landscape of the black sun. Even though he is not invoked by name, the divine blacksmith, whose belching forge is in the heart of vast, smoking, deadly Etna, is the tutelary god of twenty-first-century Southern

---

[503] Harrison, 190.

California, the beating, smoggy heart of the American military industrial complex taken to a believable condition of extremity.

In *Blade Runner*, the "nexus" generation of bio-engineered, intelligent, human-appearing "Replicants," created by Tyrell Corporation and used as slave labor "off-world," have mutinied. Replicants have been declared illegal on Earth, under penalty of death (Blade Runners are special police squads with standing orders to shoot to kill, on detection, any trespassing Replicants). These executions are called "retirement." The suggestion that "trespassing" and "retirement" are euphemisms for corporately sponsored, unrestrained police action against highly intelligent, biologically human-like slaves, creates a question from the very beginning of the film: whom are we to identify and sympathize with? What is human and what is Replicant?

A political shadow lurks in the theme of miscegenation: the threat of blending natural with unnatural creation, and the danger of ambiguous sexual boundaries. The Replicants are banned from Earth not because they are viewed as potential terrorists but because the real terror is that once released into the population they are indistinguishable from humans. The only means of detection is the "Voigt-Kampff Test," which measures minute changes in the iris in response to questions keyed to normal human emotional response—responses that Replicants, who are generated as adult forms with an artificially engineered life span of four years, theoretically have no means of developing. Like Hoffmann's feckless student Nathaniel, who goes mad from love for the fatally charming automaton Olympia, Deckerd, the Blade Runner protagonist, falls in love with a Replicant—who believes herself to be human, unaware that she is a creation of Dr. Eldon Tyrell.

Morbidly thin and frail-looking, and wearing large glasses whose lenses eerily magnify his eyes, Dr. Tyrell is head of the massive Tyrell Corporation. Like Frankenstein's monster, Tyrell's creation, the mutineering Replicant Roy Batty, seeks out his creator to demand something only Tyrell can give him. Not a mate like Frankenstein's monster (in fact, in *Blade Runner*'s steamy setting, a tribute to *film noir*'s relentless exploration of the darker side of human nature, the Replicants are the only beings who are seen to show love for each other) but a longer life span. When Tyrell refuses, Roy gently takes

Tyrell's head in his hands, places on his mouth the prodigal son's kiss and then kills him by crushing his skull and gouging his eyes.

Chew the eyemaker, whom Roy has confronted in his efforts to get to Tyrell, is an allusion to the shaggy-eyebrowed Coppelius in Michael Powell and Emeric Pressburger's 1950 film *Tales of Hoffmann*.[504] Chew's laboratory, inside what was once a strip-mall optometrist's shop—"Eye World"—an anachronistic relic in the fast-forwarded LA in which *Blade Runner* is situated, is a frozen Tartarus with icicles growing from the ceiling. The future world is so inverted that the maker's forge becomes a frozen alembic. Chew is presented as a figure dressed in a lab costume of skins reminiscent of the Arctic Tungus shaman who whimsically soliloquizes in the Asian-English-European patois which is the chaotic city's lingua franca as he tweezes a newly fabricated eyeball out of what could be an alchemical vessel. Chew is the craftsman, the technician, the artistic creator, the shaman at home with the daemonic spirits of vision. When Roy appears seeking Tyrell he tells Chew, "If only you could see what I've seen with your eyes."

Roy's eyes are an icy, celestial blue. Whereas Frankenstein's monster is ugly, plainly bearing the repugnant marks of a rogue creation, and Hoffmann's automata are beautiful but clearly inhuman to the eyes of the beholder who is not in the throes of magical enchantment, *Blade Runner*'s Replicants are beautiful and highly individualized. The humans by contrast are sweaty, harassed, chattering, atomized, and alienated.

Roy's icy fire, the fire of the fallen angels he claims as spiritual kin when he quotes Blake, is destined to be extinguished as utterly as that of Gaston Bachelard's mayfly in a candle-flame, in a single moment where "love, death and fire are united."[505] At the climax of a final battle with Deckerd the Blade Runner, Roy, his time expiring, forebears to kill his weaker opponent, even though Deckerd has killed Roy's comrades and his lover. He closes his eyes, and his death is signaled by the release of a dove from his slackened hand, which

---

[504] *The Tales of Hoffmann*, directed by Michael Powell & Emeric Pressburger (1951; New York: The Criterion Collection, 2005), DVD, in "Blade Runner Frequently Asked Questions (FAQ)," ed. Murray Chapman, 1992-1995, accessed August 24, 2003, http://www.faqs.org/faqs/movies/bladerunner-faq/.

[505] Bachelard, *Psychoanalysis* 17.

flies upward to disappear into the black sky. This is an example of what Bachelard calls the lesson of the fire. "After having gained all through skill, through love or through violence you must give up all, you must annihilate yourself."[506]

At the end of the film, Deckerd leaves the city with the Replicant Rachael. The Blade Runners will attempt to hunt them down, but the possibility of a new union, blessed by the distinctly Hermetic figure of the eccentric policeman Gaff, who appears seemingly out of nowhere upon Roy's death, signifies, perhaps, the birth of a hybrid race.

## Old and New Myths

Chris Baldick reads in *Frankenstein* an example of a purely modern myth whose main theme is distortions of power in human social relations. Baldick traces the origin of Shelley's theme to vivid metaphors of monstrosity that characterized the early debates in Britain over the French Revolution. Many writers and commentators (including Shelley's mother, Mary Wollstonecraft) passionately condemned the excesses of the Revolution and the Reign of Terror, frequently using the metaphor of "monstrousness"—regardless of which side they sympathized with—to describe the violence done to both bodies and the body politic.[507] Baldick points out that, "In modern usage 'monster' means something frighteningly unnatural or of huge dimensions." However, earlier usages of the term carry connotations that in their essence are "not physiological but moral...."[508] For example, a 1697 sermon cited in the *OED* compares the horror of "Monstrous Births" to what is termed "Monstrosity in Education." In other words, factors considered as productive of moral malaise and social disruption are metaphorically presented as a deformity of the body politic. Thomas Hobbes articulates a full sense of the metaphor of political "body" in *Leviathan* (1651):

> For by Art is created the great leviathan, called a common-wealth, or state...which is but an Artificiall Man; though of greater stature and

---

[506] Gabriele D'Annunzio, *Contemplation de la Mort*, qtd. in Gaston Bachelard, *The Psychoanalysis of Fire*, trans. Alan C. M. Ross (Boston: Beacon, 1964), 17.
[507] Baldick, 24.
[508] Ibid, 10.

strength than the Naturall.... Lastly, the Pacts and Covenants by which the parts of this Body Politique were first made, set together, and united, resemble that Fiat, or the Let us make man, pronounced by God in the Creation.509

The *Frankenstein* myth, then, reflects "the dismemberment of the old body politic as incarnated in the personal authority of late feudal and absolutist rule."[510]

Baldick identifies Frankensteinian "tales of transgression" presented within "fables of industry" not only in the works of E. T. A. Hoffmann, but also in those of Nathaniel Hawthorne, Herman Melville, Elizabeth Gaskell, and other authors fascinated by the blending of psychological and technological monstrosity.[511] Dickens too mused on artificially constructed human beings who are either "tended, cultivated or fabricated under crushing social pressures" and "end up recoiling upon their creators or parents in a predictable pattern of nemesis."[512]

Baldick notes that the significance of the specific complexities of theme in the *Frankenstein* myth becomes blurred in a persistent reductive generalization of interpretation. This reduction interprets the story as a prescient warning of the dangers of scientific hubris, and results in clumsy literalizations like the example Baldick cites from a *New York Times Magazine* story on the science of cloning from 1972: "The *Frankenstein* Myth Becomes a Reality: We Have The Awful Knowledge to Make Exact Copies of Human Beings." A subtheme of this interpretation equates *Frankenstein* with the evils of the post-Industrial Revolution machine age. As Baldick points out, such reduction "brings us right back to the cosy idealism" and the "banal liberal moral" offered by many writers. Baldick offers the example of one writer whose musings on *Frankenstein* concluded with the observation that:

> our technology reflects ourselves and our motives; if we replace repression, fear, arrogance, and the desire for control with acceptance, love, humility, and the capacity for understanding, our machines can become a well-adjusted part of our culture.[513]

---

[509] Thomas Hobbes, qtd. in Baldick, 15.

[510] Baldick, 16.

[511] Ibid, 63.

[512] Ibid, 111-112.

Donna J. Haraway introduces the term "cyborg feminism." Her purpose is to re-appropriate cultural terms for those excluded by the "master narratives deeply indebted to racism and colonialism."[514] Among these master narratives is the political appropriation of the terms *nature, art* and *technology*. Haraway's cyborg is a hybrid creature, composed of organism and machine, "a creature of social reality as well as a creature of fiction."[515] In short, a Hephaistean creature. As the choke-hold on power of the "master narratives" weakens, Haraway's "Cyborg Manifesto" looks forward to a world which "might be about lived social and bodily realities in which people are not afraid of their joint kinship with animals and machines, not afraid of permanently partial identities and contradictory standpoints."[516] In such a fruitfully ambiguous world, artists, whether male or female, would be responsible for their own narratives and politically empowered by what depth psychologists like Stein have termed "wounds." Haraway's cyborgs are:

> odd boundary creatures…all of which have had a destabilizing place in the great Western evolutionary, technological and biological narratives. These boundary creatures are, literally, *monsters*, a word that shares more than its root with the word, to *demonstrate*. Monsters signify.[517]

Examples of "monstrous" boundary art include Austrian performance artist Stelarc's "Pingbody" events which allow remote audiences to "view and actuate Stelarc's body via a computer-interfaced muscle-stimulation system," such that his body's "proprioception and musculature is stimulated not by its internal nervous system but by the external ebb and flow of data."[518] "Biotech" artist Eduardo Kac's project "The Eighth Day" presents "a self-contained ecological system" effected by means of genetic editing or even the introduction of "'artist's genes,' i.e., chimeric

---

[513] Baldick, 7-8.

[514] Donna J. Haraway, *Simians, Cyborgs, and Women: The Reinvention of Nature* (New York: Routledge, 1991), 1.

[515] Ibid, 149.

[516] Ibid, 154.

[517] Ibid, 2.

[518] Stelarc, "Pingbody; An Internet Actuated and Uploaded Performance," 1996. Accessed November 23, 2004. http://www.stelarc.va.com.au/pingbody/index.html.

genes or new genetic information completely created by the artist" resulting in transgenic creatures (such as plants, amoebae fish and mice).[519] Artists who occupy Haraway's boundary territories are pushing the boundaries further through exploration of Hephaistean questions concerning the mutual limits of science and culture, technology and art.

## God of the Military-Industrial Complex

In his farewell speech as president in 1960, Dwight D. Eisenhower gave the following warning:

> In the councils of government, we must guard against the acquisition of unwarranted influence, whether sought or unsought, by the military-industrial complex. The potential for the disastrous rise of misplaced power exists and will persist.[520]

Eisenhower warned that, "Only an alert and knowledgeable citizenry can compel the proper meshing of the huge industrial and military machinery of defense with our peaceful methods and goals, so that security and liberty may prosper together."[521]

David L. Miller identifies Hera, Prometheus, and Hephaistos as gods of the military-industrial complex.[522] Eisenhower's speech gives the reasons why Hera, whose interest is in power and its perquisites, would come at the front of Miller's list. Homer's *Iliad* paints a picture of Hera's attempts to wield power in behalf of those she favors and destroy those who have slighted or offended her. Eisenhower also asked Americans to be attentive to the seductive pull of innovation for its own sake: "Akin to, and largely responsible for the sweeping changes in our industrial-military posture, has been the technological revolution during recent decades."[523] In this we may

---

[519] Melentie Pandilovski, "The Ghost of Biotechnology: Art of the Biotech Era," February 2004, par. 19, The Experimental Art Foundation and Dark Horsey Bookshop, 2004, accessed November 10, 2004, http://www.eaf.asn.au/essay.htm; Eduardo Kac, "Transgenic Art," par. 7, KAC, accessed November 10, 2004, http://www.ekac.org/transgenic.html.

[520] Dwight D. Eisenhower, "Military-Industrial Complex Speech," sec. IV par. 3, Public Papers of the Presidents, Dwight D. Eisenhower, 1960, accessed October 30, 2004, http://coursesa.matrix.msu.edu/~hst306/documents/indust.html.

[521] Ibid, sec. IV par. 6.

[522] David L. Miller, *The New Polytheism: Rebirth of the Gods and Goddesses*, (New York: Harper, 1974), 66-67.

see Prometheus as the rash innovator who "ends trapped on a rock, gnawed at by the power he has himself supplanted by his knowledge." But why Hephaistos? I do not agree that the answer is because he "is at a total loss for sensuousness and feeling."[524] Here also, Eisenhower offers food for thought: "Today, the solitary inventor, tinkering in his shop, has been overshadowed by task forces of scientists in laboratories and testing fields."[525] Although the smith-gods and the laboratory scientists seldom wield sole temporal power, their making is nevertheless implicated in it, as Hephaistos is when he answers Hera's summons to fire on Xanthus or when he performs a good job of work, however unwillingly, of chaining Prometheus to the rock where he will be tortured with unbreakable chains only Hephaistos can fabricate, at the command of Zeus.

Two twentieth-century fictional works in which Hephaistos is mentioned by name are the 1989 film *The Adventures of Baron Munchausen* and Sten Nadolny's 1994 novel *The God of Impertinence*, in which Hephaistos is a central character. Terry Gilliam, director of the film, makes use of the version of the Munchausen story published in 1793, which features, among the Baron's many adventures, a visit to Vulcan's workshop in Etna.[526] The original Munchausen view of Vulcan, which Gilliam honors, is already an Enlightenment one that includes the notion of hierarchical management, i.e., Vulcan as managing director of the forge. The Baron expresses his astonishment to find himself, fallen into Etna, "in the company of Vulcan and his Cyclops, who had been quarrelling, for the three weeks before mentioned, about the observation of good order and due subordination...."[527] In short, the Baron interrupts a labor dispute. In the 18th century version of the Munchausen story, these quarrels are experienced by terrestrial dwellers as volcanic

---

[523] Eisenhower, sec. IV par. 7.

[524] Miller, *Polytheism*, 67.

[525] Eisenhower, sec. IV par. 9.

[526] *The Adventures of Baron Munchausen*, directed by Terry Gilliam (1989; Culver City: CA, Sony Pictures Home Entertainment, 1999), DVD.

[527] "The Travels And Surprising Adventures Of Baron Munchausen Illustrated With Thirty-Seven Curious Engravings From The Baron's Own Designs And Five Illustrations By G. Cruikshank,"1793, ch. 20, par. 3, ed. Marcus Rowland, 2002, Forgotten Futures, accessed August 24, 2002, http://homepage.ntlworld.com/forgottenfutures/munch/munch.htm.

eruptions. Gilliam's version of Baron Munchausen's adventures is updated with a signal, anticipative anachronism: when giving Munchausen a tour of his cavernous forge, a sweating Vulcan displays with pride to a bemused Baron "our prototype," a nuclear ballistic missile. The bored Baron, waving a lace-fringed cuff, asks what it does. "It kills the enemy. All of them. All their wives and all their children and all their sheep and all their cattle and all their cats and dogs." Moreover, this can be done from a distance, simply by pressing a button.

In *The God of Impertinence*, Sten Nadolny spins a near-future tale in which Hephaistos has become the Lord of the Universe in a world driven by computers—very much the world we live in. Nadolny's novel turns on his mythopoesis of Hephaistos as the present-day cigar-chomping supreme Manager who manipulates the global economy from behind the scenes. Hermes, who enters the story after having been chained in a defunct volcano for 2,187 years and who has been newly—and mysteriously—released, quickly sees that,

> Hephaestus' power had seemingly increased.... No wonder—he was always making something (it was his form of eroticism), and unless someone stopped him, it was only a matter of time till that unbridled demiurge was omnipresent through his artifacts alone.[528]

In his novel, Nadolny introduces a theme he calls *multiplication*, "crucial to everyone's happiness" which "consisted of frantically increasing identical elements in a slavishly reproductive drill."[529] This he attributes to Hephaistos, the wounded archetype of the maker gone very far awry.

Nadolny's Hephaistos plans to manipulate human will to accomplish "pushing the button." The resulting holocaust will obliterate not only humankind but release Hephaistos from his loveless condition and the rest of the deathless gods from their terminal boredom. Yet it is Hephaistos' (and Nadolny's) native artist-wisdom that causes him to recognize that Hermes, whom Hephaistos had chained, needs to be liberated, to bring needed revolution to a literalist world deprived of mystery, in short, to re-spark the fire.

---

[528] Sten Nadolny, *The God of Impertinence*, trans. Breon Mitchell (New York: Viking/Penguin, 1997), 34.
[529] Ibid, 35.

Hephaistos himself recognizes that without Hermes' mercurial ambivalence, his *techné* becomes stale and death-serving, and it is Hermes who duly awakens Hephaistos to love, thereby saving the world from immolation. Is Hermes intuitively aware of Hephaistos' repressed access to *charis*—beauty, love, and grace? Are we aware of our own? (And, what is it in the Hephaistos archetype that calls forth the Hermetic?) Indeed, seeing the peculiar, dandyish character Gaff with his ever-present silver-tipped cane (or, Hermetic staff) through the Hermetic lens also helps make sense of the mysterious, hopeful sparks of light *Blade Runner* offers, even in the midst of overwhelming darkness.

The Hephaistos imagined in connection with the military-industrial complex does not question the status quo, following orders when summoned from above, quietly going about his own underground business in the meantime—or, indeed giving orders that allow the machinery of commerce to grind on, even when Zeus can't care less. The industrial giants who profit from war certainly have a large share of Hephaistean fire. 'It's only a product and we're making an honest living.' As Vulcan confides to Baron Munchausen, "Oh, we cater for all sorts here. You'd be surprised!"[530]

Interestingly, the so-called "war cabinet" of the Bush Administration (Dick Cheney, Donald Rumsfeld, Colin Powell, Paul Wolfowitz, Richard Armitage, and Condoleeza Rice) are known to be friends of long standing who have met for years to refine their ideological position and formed a sort of self-described "club" to which Rice gave the name "The Vulcans." Rice is a native of Birmingham, Alabama, where a "mammoth fifty-six foot statue of Vulcan on a hill overlooking downtown paid homage to the city's steel industry." The cast-iron statue was originally commissioned in 1903 for the St. Louis World's Fair. The god is depicted as brawny, with muscular chest, arms, and legs and wearing (only) a smith's apron and sandals. He holds a spear point in an upraised hand; the other, holding his hammer, rests on the anvil. According to James Mann, author of *Rise of the Vulcans*, "That word, *Vulcans*, captured perfectly the image the Bush foreign policy team sought to convey, a sense of power, toughness, resilience and durability." Mann points

---

[530] *The Adventures of Baron Munchausen*, dir. Terry Gilliam.

out parenthetically that the statue "was taken down for repairs in 1999 because it was beginning to fall apart."[531] He implies that this fact symbolizes the unconscious shadow side of the group's identification with the god, since many see the their policies as conveying the Vulcans' sense of aggrandizement, closed-mindedness, rigidity, and brittleness. What fell apart, clearly, was a myth that could not be seen in its psychic reality.

Long before its dismantlement, the statue endured increasing indignities—it became a billboard for selling soft drinks and pickles. At some point the sword point was exchanged for a light bulb that remained illuminated for a week whenever a traffic fatality occurred in the city. As it so happens, since Mann's book went to press the statue has been re-erected, having been lovingly restored to its original condition as a hopeful emblem of the revitalization of Birmingham's steel industry. An article about the statue's history posted on a local booster website reports that "his bare derrière still faces a suburb to the south—a feature known locally as 'Moon over Homewood.'"[532] Suffice it to say that Hephaistos lives.

But, Hephaistos lives in the midst of our culture in a literalized and diminished state. Yet, surely there is significance in the re-rising of Homewood's "moon" over Birmingham. How do we discover that significance?

## *Whither Hephaistos?*

As I have demonstrated, the myth of Hephaistos contains considerable energies across a particular spectrum of divine and human activity. He is the god of making and of working transformation on physical matter. The lineage of the blacksmith gods and the chthonic phallic cohort represents the maintenance of creation, which by necessity involves the ability to hold opposing forces in creative tension. What would be the effects of regaining

---

[531] James Mann, *Rise of the Vulcans: The History of Bush's War Cabinet* (New York: Viking, 2004), x.
[532] James Book, "Return of a Giant: A fully restored Vulcan—Birmingham's 100-year-old statue—resumes its rightful place in town." About Vulcan (pdf), par. 14, Vulcan Run 10K and Birmingham Track Club, 2004, accessed October 30, 2004, http://www.vulcanrun.com/.

aspects of the mythos of the god that are now found only at the unconscious level of the cultural mind? How might these mythic strands be regained?

Bachelard gives an example of what he calls a *"diminished legend"* concerning the old English stories of Wayland the Blacksmith. Wayland is a "subterranean power," in the lineage of the mountain dwarf and smith. A passing knight has only to tether his horse to "a large rock known to every inhabitant of the nearest village," leave money, and depart, whereupon he would hear "the sound of hammer on anvil" and return to find the money gone and his horse freshly shod. Sir Walter Scott dealt with this material in his novel *Kenilworth*, but, Bachelard observes, turned an "obscure legend, better revived through a long meditation on the subterranean powers, into a casual whodunit" that rationalizes the legendary smith into a wretch with a checkered past who resorts to exploiting a quaint local legend as a clandestine way of earning his living from the "ignorant boors" of the neighborhood.

Bachelard observes that "A similar diminution of legend can be seen in works which pretend to exalt *the powers of legend*. This *deaf rationalism* kills mythic *élan*." Bachelard offers the example of Richard Wagner's Siegfried, who reforges the broken sword of his father Siegmund—which the dwarf-smith Mime cannot—by following a dream that inspires him with the idea of filing the sword to dust. He then places it into a form to remold it, to Mime's great astonishment. What Wagner has left out of his opera from the original mythic story is that the legendary smith Wayland took the sword dust, mixed it with flour and milk to form a paste, which he then mixed with poultry fodder—and forged the droppings of the chickens. The metal, flour, and milk thus undergo a biochemical baking, an oneiric image that requires, Bachelard notes, "A great deal of long and careful reverie" to make poetic sense of it. One can see Wagner's literalizing in this example, though Bachelard offers Wagner the benefit of the doubt by pointing out that contemporary mythologists writing on the image of "bits of iron swallowed by fowl" declared it "meaningless" within the context of the legend, though they also went on to report that this mytheme "can be found along the Rhine and the Euphrates" (Bachelard expresses rhetorical shock at the "large and faithful audience for a 'meaningless' storyline!").[533] Indeed, this image gives mythic weight and significance to the sense in which Bachelard's reverie as dreaming-into image, requires *rumination, digestion,* in the sense of the slow cooking of somatic experience. In this sense, the whole body and being is inducted into reverie and the image is incorporated somatically.

Seen from this perspective, it becomes far harder to make Hephaistos' chronic lameness a pathology to be diagnosed and set aright so that he can walk straight in the ways of the patriarchy. Delcourt asserts that Hephaistos' fall results both in his mutilation and his initiation into magical powers.[534] Initiation comes with and through the experience of the archetypal condition of crippling, woundedness, blindness. It does us little good, then, to prescribe to Murray Stein's "Hephaistean man" a remedy to heal his woundedness. The god who deals creatively and transformatively with matter is a limping god. What does this mean?

For Hillman, the rationalizing tendency is a function of the literalism that is the effect of being locked into the ego's view. This follows from his insight that the symptom—the wound—is an expression of the soul's autonomy. "Symptoms remind us of the autonomy of the complexes; they refuse to submit to the ego's view of a unified person." In other words, we are not in control of the wounded or sickened state—who would consciously wish impairment upon himself?—for which reason the symptom "gives me the sense of being…in the hands of the Gods."[535]

Hillman also engages the idea of beauty as soul-making. By "beauty" he does not mean merely the aesthetic aspect of appearance. Beauty is *substance,* an *"epistemological* necessity" whereby the gods attract us into life by touching us through our senses. "As well, beauty is an *ontological* necessity" that grounds the world in its sensate particulars. Aphrodite's golden nakedness is soul's shining-forth in the capacity of the heart to engage with individual events amid the swirling appearances of the cosmos. The heart is that organ of what the Greeks termed *aesthesis,* "sense perception," from which

---

[533] Bachelard, *Earth*, 129.
[534] Delcourt, 119.
[535] James Hillman, *Re-Visioning Psychology* (New York: Harper, 1976), 48-49.

we derive "aesthetics"—through which I separate the special arch of my lover's eyebrow from all other phenomena. Philosophy that resorts to abstractions to make sense of appearances is distorted from "its true base," says Hillman. For, as:

> philosophy takes rise in *philos* [love], it also refers to Aphrodite in another way. For *sophia* originally means the skill of the craftsman, the carpenter (*Iliad* XV, 412), the seafarer (*Hesiod, Works*: 651), the sculptor (Aristotle, *Nichomachean Ethics*, vi:1141a). Sophia originates in and refers to the aesthetic hands of Daedalus and Hephaestus, who was of course conjoined with Aphrodite and so is inherent to her nature.[536]

Making, then, must be understood as an avenue to the soul, and art-making as both an oneiric and a practical activity whose transforming effects may be felt in the "real" world.

## Talking with Gods

Mary Watkins summons the observation by Reinhold Niebuhr that, "in the Hebraic tradition human beings were distinguished from all other living creatures not by virtue of their capacity for reason but by virtue of their engagement in three kinds of dialogues: dialogues with neighbors, with themselves, and with God."[537] The last two, observes Watkins, have become discredited to the degree that revaluing imaginal experience requires letting go of the prevailing societal norm. This norm favors a notion of "development" that is founded on the patriarchal norm that assumes rational thought and social intercourse are paramount to mature human effectiveness and identity. The internally-focused dialogue with an unseen other, which during childhood is often audibly voiced, is thought to decline and diminish with "normal" development, to be superseded by "the relative silence of abstract thinking."[538] What we come to think of as "daydreams" come to be seen by adulthood as "an escape from the demands of consensual validation of reality."[539]

---

[536] Hillman, *Blue Fire*, 302.

[537] Mary Watkins, *Invisible Guests: The Development of Imaginal Dialogues* (Dallas: *Spring*, 2000), 1.

[538] Watkins, 40.

[539] Ibid, 38.

From Watkins' point of view, Plato depicts the internal dialogue as fundamental to reason. In the *Greater Hippias*, Socrates says that when he goes home at night, he returns to the company of a voice. This voice, as Watkins quotes Hannah Arendt as commenting, is "a very obnoxious fellow who always cross-examines him." Says Socrates of this 'fellow,' "he asks me whether I am not ashamed of my audacity in talking about a beautiful way of life, when questioning makes it evident that I do not even know the meaning of the word 'beauty.'" By contrast, Hippias remains by himself at night, for, in Arendt's words again, "he does not seek to keep himself company. He certainly does not lose consciousness; he is simply not in the habit of actualizing it." Unlike Socrates, Hippias ceases to think; he refuses to open a dialogue.[540]

Imaginal dialogues, continues Watkins, represent:

> a breach with a secular view of reality which holds that one's conversations are not to be peopled by gods, angels, muses, gnomes, or other strange characters. They also constitute a breach with a unitary concept of self that relies on a stable identity and does not look closely at shifts of mood, tone, or attitude that might suggest a multiplicity of the self.[541]

Hillman asserts that a polytheistic psychology that allows for these dialogues "can give sacred differentiation to our psychic turmoil and can welcome its outlandish individuality in terms of classical patterns."[542] These patterns are the gods and goddesses, and our conversation with them may be seen as *dialogue not pathology.*

## Re-Mything Art and Technology

Recognizing Hephaistos in images of art and technology that reflect mainly pathology, malaise, and potential holocaust suggests that new images must be found in which we may recover the *mythos* of the limping god of fire and of making and of creativity, images that reflect the material well-being the Homeric Hymn tells us comes

---

[540] Ibid, 49-50.

[541] Ibid, 45.

[542] Hillman, James, "Psychology: Monotheistic or Polytheistic?" *Working with Images: The Theoretical Base of Archetypal Psychology*, ed. Benjamin Sells (Woodstock: Spring, 2000), 40.

from Hephaistos and that come into constructive dialogue with the archetypal energies the god represents.

Art critic and painter Suzi Gablik believes the choices for artists in contemporary society are two: to remain in the state of what Murray Stein has described as "woundedness" and Gablik sees as a condition of "extraordinarily paralyzing, cynical alienation" that can only seek to deconstruct society's malaise, or, to choose "'the new common sense' of the pragmatic visionary."[543] Gablik articulates the need for a "reconstructive postmodernism" whose most powerful aspect is healing:

> Increasingly, as artists begin to question their responsibility and perceive that "success" in capitalist, patriarchal terms may not be the enlightened path to the future, which of these views they hold definitely affects how they see their role: as demystifier or cultural healer.[544]

Gablik offers many examples of the latter, including New York teacher and artist Tim Rollins, founder, together with a small band of his students at Intermediate School 52 in the South Bronx, of K.O.S. ("Kids of Survival"). Many of Rollins' students have been "classified as emotionally handicapped and learning disabled." Rollins was dissatisfied with the separation between teaching and making art and decided to combine the two in a way that engaged his students as partners.

> "Because many of them were dyslexic," he says, "I would read to them, not those embarrassing primers, but books like *Frankenstein*, Dickens' *Hard Times*, Kafka's *Amerika* and Dante's *Inferno*. The kids went crazy for them. I would have them draw while I read, but it wasn't that I wanted them to illustrate what I was reading. Instead, I told them to come up with visual correspondences between the stories and things in their daily lives."[545]

Instead of working to correct "disability" Rollins' students worked with the realities of their own lives, creating art based on a dialogue with texts, each other, and the medium of art. Many of their art works are in major collections, but more importantly, Rollins and K.O.S. travel and conduct workshops in schools and institutions.

---

[543] Suzi Gablik, *The Reenchantment of Art* (New York: Thames and Hudson, 1992), 25.
[544] Ibid, 23.
[545] Gablik, 106.

They are carrying on the work of re-engaging a generation of kids in re-mything the canon of literature by discovering its meaning for them. No pathology, no therapy; rather, art-making, through a process of dialogue with imaginal and real others. In dialogue with Aeschylus' *Prometheus Bound*, K.O.S. identified a key question with meaning for them: "What is this fire?" This "fire" is, "Creation, warmth, making," "Potential, possibility."[546] K.O.S. member Daniel says, "Prometheus and Hephaistos are like brothers.... They understand each other because they do the same thing. They create. They're artists. They both know how to harness fire to make something good."[547] K.O.S. member Emanuel says, "We want to create because we don't just want to sit back and take whatever someone else dishes out to us. We need to make something new…to make something out of our lives."[548]

As for re-mything Hephaistos the infernal technologist, consider as an example the work of William McDonough and Michael Braungart, an architect and chemist respectively, who have teamed to create what they term "Cradle to Cradle Design," which uses strategies they call "eco-effective" in order "to create products and systems that contribute to economic, social, and environmental prosperity."[549] McDonough and Braungart's model for industrial and architectural sustainability envisions industrial-strength making without negative impact on the environment and literally without waste of any kind. The vision is not merely to reduce environmental harm from industry, but to create positive effects.

McDonough and Braungart have created collaborative projects with a list of companies that includes Nike, Volvo, BP and BASF. A key project of McDonough and Braungart's firm, MBDC, is being carried out in collaboration with Ford Motor Company. Ford's historic Rouge River manufacturing plant in Dearborn, Michigan was hailed as, "A marvel of modern engineering" that at the time of its

---

[546] Tim Rollins and K.O.S, "Prometheus Bound: A work of art for the web," par. 38-40, Dia Art Foundation 1995-2004, accessed November 6, 2004, http://www.diacenter.org/kos.

[547] Ibid, par. 37.

[548] Ibid, par. 41.

[549] MBDC, "Firm Profile," MBDC, LLC., 2001, accessed November 6, 2004, http://www.mbdc.com/firm_profile.htm.

heyday in the mid 1930's, employed 100,000 workers.[550] Raw materials entered at one end of the plant and were converted to completed automobiles exiting at the other, creating an enduring manufacturing standard. The renovation project will seek to optimize assembly operations using the newest manufacturing technologies and processes designed to dramatically reduce all waste. Using the "Cradle to Cradle" lens, Ford's plans for the site will in effect place the manufacturing activity in ongoing dialogue with the natural environment surrounding it. The Rouge area watershed will be protected from storm runoff from the huge plant by means of an ecologically designed 500,000-square-foot roof—the largest of its kind—which will hold and recycle several inches of rainfall. Plant species capable of absorbing contaminants that were by-products of the plant's past operations will be used to restore the soil by Phytoremediation. Energy use will be reduced by using trellises for flowering vines and other plants to shade and help cool a new office building and the refurbished assembly plant, and through the use of solar cells and fuel cells. To restore the plant area as a wildlife habitat, 1,500 trees and thousands of other plantings will attract songbirds and create habitats.[551]

The renovation is slated for completion in 2020. A massive manufacturing plant will always affect the environment. It is a deformation of the natural scene. However, instead of destroying it or throwing it away, it can instead generate a transmutation in its surroundings, through creative thinking and making. Ford is also working on its "Model U" automobile—conceived to be as revolutionary as the "Model T" was a century ago—which will incorporate design principles and materials that are biologically safe and environmentally positive. Ultimately, Ford will recycle its customers' used cars.[552] "Cradle-to-cradle" means no waste, as we think of it now, at all.

---

[550] Ford Motor Company, "Rouge Renovation," 2004, par. 2, accessed November 6, 2004, http://www.ford.com/en/goodWorks/environment/cleanerManufacturing/rougeRenovation.htm.

[551] Ibid, par. 6.

[552] See: MBDC, "Monthly Feature: February 2003, A Model for Change," MBDC, LLC., 2003, accessed November 6, 2004, http://www.mbdc.com/features/feature_feb2003.htm.

The examples above must be understood as belonging, at least in part, to the mythos of Hephaistos. And, if a single example of Hephaistean technology already exists, it must be the Internet: the fantastic product of *technê* and *mêtis* together, the magically crafted, invisible net that now covers the globe, providing links for commerce as well as subversion; for communication of all kinds, containing all possibilities.

The god and his myth are far from dead. In fact, the examples above represent a mythopoesis—the mechanism whereby myth is continually revitalized—of the Homeric Hymn:

> Sing, you clear-voiced Muse, of Hephaistos renowned for craft,
> Who with bright-eyed Athena taught splendid works to humans on earth—
> They had before then been dwelling in caves on the mountains like beasts,
> But now, knowing works through Hephaistos renowned for his skill, with ease
> Till the year brings its end they live in comfort within their own homes.
> Come now, be kindly, Hephaistos; grant us prowess and wealth.[553]

---

[553] "Hymn to Hephaistos," trans. Crudden, 84-85.

CHAPTER 6

# The Last Word on Recovering the Mythos of the God: Hephaistos Speaks

HEPHAISTOS, that's me. I have a reputation for being the ugliest Olympian. In fact, the only ugly Olympian. My mother Hera, the hag, cast me off Olympus when I was born. Ugly, lame. Whose fault might that have been? What did she do to bring down such a fate? As she did not solicit the participation of the great Zeus in my conception, she could not blame him when it went wrong. An infant, I fell from heaven to earth. By the time I landed on Thetis's shore I no longer even knew fear. I have flown. But I am not a creature of the aether. Not that way.

I have a forge, and I have skill, and my talent is everything that is godly about me. Solid ground is my element; that, and fire, and all the noble metals, especially gold and well-wrought bronze. Both gods and men know well what is truly of value. That which is solid and well made. Even my bitch of a mother knows where to find me when she hears of a new device I have made. Unlike some of the others, she is less interested in personal adornment. Everyone in all the worlds knows I have no equal as a maker of ingenious and exquisite objects. Who else but I could cover Aphrodite in well-wrought gold and gems, to make her charms more resplendent, veiled or no. Of course she rarely is, veiled that is. And who of any intelligence whatever would expect the Goddess of Love to stay in one bed, whether god's or mortal's? Marriage has ever been an economic institution. My mother, who knows it well, would pretend otherwise, but this is only pretense. Did I not receive Aphrodite for wife as payment for freeing Hera from her sticky throne? The trick I later played on that adulterous mare, Aphrodite, breathless in her lust after that lunkhead Ares with his brainless head and straight legs and back—well-fashioned, I grant—was out of pique, this is quite true. Not even I could maintain my even temper at some of Aphrodite's more blatant jibes. So, I divorced her. What no one knows is that the next time she came to me, smiling, knowing I had fashioned another jewel, more artful even than the rest, well....We both know what has value and keeps it.

My mother is a curious blend of pretense and stolid powerful anger. I know what I took from her at my birth. She, of all the gods, is least interested in display, except of a courtly kind to maintain her queenly image of propriety, which only thinly veils her raw lust for power and intrigue. What she seeks me out for most keenly are the devices only I have the wit and skill to create, and are useful to her in certain of her intrigues. Devices that bring her intelligence of things, like Talos, the man I made of bronze whom Zeus gave to Minos. Minos sent him to walk every road in Crete to keep the king informed of hidden plots and misdeeds. Or the occasional implement to be used on a careless nymph who, warmed by the caresses of Zeus, has grown too bold; or any demi-god or mortal like Ixion, foolish enough to insult her, usually out of their own blundering ignorance.

We get along, Hera and I, the old bat. There are few who do not know that her deathless beauty does not mirror what is underneath, though she is far more complex than they guess, but then, that's true of all the gods. I tricked her—it wasn't hard to do—when she sat in that golden throne I sent her, the one that imprisoned her and held her in midair for all to ridicule, though I know they were shocked and afraid. She could not resist its regal beauty. No trick jewel would have worked as well as the emblem of royal power to stir her lust and make her uncautious. What a mother, who after throwing away her son, could convince herself he would love her enough to send such a gift out of pure filial love and respect? I don't mind in the end; I have created myself out of fire and bronze and sheer talent. I cannot begrudge her forever, there's no point to it.

Zeus certainly called for me double-quick, suddenly wanted me, his "son" on the mountaintop among the deathless gods who are my kin, for they soon found only I could free his queen. Thus he claimed me. But I wouldn't come. He sent Ares after me and I singed him! Ha! He shot away double-quick from the flames I aimed at him. I can congratulate myself for overwhelming the god of war with sheer force, no question. I can summon it at will, as I did in Troy when Hera sent me against the mighty river Xanthus, who begged for mercy when his banks shrunk and waters sizzled at my mighty flames. He renounced Troy right away, he who had been one of their greatest allies. He laughed first, when he stole Achilles, but had to give him up and did not end laughing. For me it was no great matter, a job like any other.

After Ares came hot-footing it back to Olympus, then Zeus, who is subtle, I'll give him that, sent Dionysus to sweet-talk me. Who could wonder that I enjoy subtlety and the lure of the senses and honeyed talk and the well-deserved praise I receive when they see my work? And he flattered me greatly, admired everything in the workshop. He has taste. Ah, but it was the wine that did it. Wonderful stuff! And the technology to make it is a great wonder for which I am even now filled with admiration, as I of course admire all well-wrought things. Of course I avoid it now. I entered Olympus on the back of a mule and not riding nobly astride either. The deathless gods, deathless like me, split their sides laughing. I learned an important lesson then I've never forgotten. They had to respect me anyway, because my skill is so great, no one but me could free the Queen of Heaven from her unfortunate throne. But they hide their awe of me in laughter. It's useful.

They laughed when I caught beautiful Aphrodite, that whoring mare, and idiot Ares who lacks all subtlety, under my invisible net. I got over my anger in time. I have better things to do than hold grudges. And let them laugh. By playing the clown I can achieve many things amid the crowd of my imperious kin on Olympus. Hermes knows as much, the liar. Who else besides me but Hermes himself could ridicule the posturing Ganymede and get away with it? The gods knew well they laughed not at my shuffling gait alone. Ha!

But why should I care? Many there are who know what is truly valuable but will not say so openly.

It is, however, the blaze of the shining, precious metal that attracts them all to me and keeps them coming back, full of awe and greed for the beautiful things. I have no need for mirrors. My mirror is the work I do. And I see myself well enough reflected in the light dancing in a nymph's eye when she fingers an exquisite curving bracelet. Ah, if only you knew how many of them know well the connection between an eye for beauty and a steady hand, and the aesthetic pleasures to be had outside the forge when the fires are banked and the anvil stowed. Think you: did not the figures I crafted on Achilles' shield dance, their well-wrought limbs shining and joyful? Were all who marveled at it not convinced of the astonishing naturalness of it all, how they seemed filled with the fire of life itself, more than just the hardened surface left by the fire's heat? I have flown, and I can far more easily imagine myself into the stretching limbs straining with joy, more easily by far than those of you who are whole-bodied and smug would ever stop to think. It's merely the pleasurable exercise of imagination, and the fruits of all the time I had in Thetis's cavern—where for nine years the goddesses of the sea hid me from god and mortal alike. There, every day, I watched the leaping, sinuous dolphins with their bright glinting eyes (and often heard my foster-mothers conversing with them), and chased my crab-brothers who walk like I do and have bright carapaces that shine like hot metal. All the intelligent creatures have their subtle habits and their particular beauty, each different from the other—can you not see it—and at the same time all coming into their proper forms according to the character of their kind and the mystery of structure that seems to lie inside all creatures, giving even the gods their forms. I think even Zeus has little ken of this. I had time and training, which I used well, and whatever Hera may have done amiss to abort me less than perfectly formed, I was born with a genius for seeing and for making. And who with a little imagination would not think that lovemaking too is part of all that? (And who, lacking in imagination, is worth having as a lover?)

For Zeus I have admiration. True, he threw me from Olympus a second time—for an earth-bound god I have flown far and long since my birth—but why should he not? I took my mother's part that one time she tried to overthrow the king of heaven. He has a mighty temper, and even though I perform tasks for my mother and see fit to do as she bids, it is well to have the Thunderer to rule the unruly in heaven and the worlds below. I have no dislike for order, and so long as my work pleases the gods, as it ever shall, they leave me in peace to go about it. And, someone has to work. They laugh and scorn work, but I have my graceful, well-formed wife Aglaea, gift of her mother, Eurynome, to charm my guests and cheer me in the short time I spend outside my forge, and mind accounts. It is a good way to spend immortality. I am never bored.

Considering I was the disowned one, I have much family. At least three mothers: my loving foster mothers Eurynome the ancient one, and Thetis the Nereid took me in, healed my wounds and gave me a tutor, Kedalion of Naxos, who initiated me early, as is proper and necessary, into the secrets of the forge. I was happy to drop everything to fashion armor for Thetis's son Achilles at her urgent request, though we all knew even that glorious armor, which came to be famed throughout the world, could not prevent his ordained death on the plains of Troy, or the grief of my dear foster-mother, even fore-knowing of his death as she did. But did she not cajole great Zeus, grasping his knees, to turn the tide of war against his own purposes—and worse, against Hera's—so that the Achaeans had to beg Achilles, the best of them all, to fight, that he might thereby die a hero? And wasn't my mother Hera angry! And how Zeus raged at her, nettled, I'm sure that he had been seen promising Thetis such a favor; beautiful Thetis, whom they call "Silver-Footed." She did not have to beg; Zeus owes her far more than any want to admit. For, it was her alarm that brought the powerful Hundred-Hander to rescue Zeus when his brother Poseidon and sister Hera—for she was his sister and his equal before she was his wife—and even, they say, Athena, almost succeeded in binding him up for good. A strange crew, I doubt they would have succeeded in ruling Olympus without tearing each other to pieces and the whole thing reverting to Chaos! But I have learned much from my foster-mother of holding one's power discreetly, making peace prudently, and taking action when necessary. Like hers, few guess my powers.

When Zeus raged back at my mother that day in the banquet hall, threatening to throttle her and everyone else, none of the other gods raised a finger, all quaking in their boots. Only I dared to step in. Not to threaten openly—though I'm far bigger and brawnier now than when Zeus threw me from Olympus the last time (of course it's my work that does that, but it's not my force I'm praised for by anyone who remembers to praise me: it's my cleverness). It wouldn't have worked anyway. But what I did was to take them all off their guard by clowning. I even got to remind Zeus to his face of his despicable treatment of his "son" that last time, even while I was calming the lot of them down. And I got away with it, so seductive were my *logoi*, my honeyed words, skillfully chosen. The rest of the gods forgot their fear, the cowards, and laughed, nervously at first of course. But now, even my mother looks on me with admiration, I whom she despised at birth. What matter? She listens to me now.

Father? It seems convenient for Zeus to claim me as his own, so father have I, after a fashion. He was angry that once, but he has much need of me, to build his mansions of state and forge his mighty thunderbolts. He needed me when he wanted the woman Pandora made from scratch, and it pleased me to send her off with the most beautiful crown that I ever made, wrought in gold, with all the charming creatures of Earth on it, more beautiful than any I had made for any goddess. I hoped it would give lovely Pandora some pleasure in her new home on earth among men, and show them what a gift they'd really gotten. It seems to me beauty is too little appreciated for its own sake. It was a dirty trick though, and by the time Zeus sent me in the company of his two thugs to chain Prometheus whom he accused of starting it all in the first place, my patience was wearing thin.

But I am not always on Olympus. I have other business to attend to. My people in Lemnos, whom I love as they love me, are always glad to see me, and I have clients far and near, like Alcinous, whose palace is the most beautiful in the world. I made it, of course. And I visit my brother-smiths in Egypt and in Crete, though less now than formerly, in earlier ages. Lately I spend more time in Sicily. Of course I keep track of the new initiates among the sons of Kabeiro, for the art must be passed on to new generations (and, though few are aware of it, I enjoy a marital relation with this august goddess as well, the mother of her people, and I father of her sons). Neither gods nor mortals can prosper without the gifts of the daemons of the bright metals, or indeed that dull metal, iron, the strongest of all. These are all activities I keep to myself. It is the nature of the art to be secret and hidden, and there is much more to it than the glamorous blaze of the forge-fire, so fascinating and so frightening to mortals and so necessary to the gods. (The gods seem sometimes to forget that if I am not there, there will be no fire to carry their sacrifices to heaven on its smoke.) And there is more than just fire: the daemon of the bellows is the same as that which drives the winds, which you feel now as a blast, fit to enrage iron so that it turns to water, with enough power to fell trees; now as a tickling breeze that teases the most delicate gold wire into shapeliness or gently nuzzles the hairs on your skin. Water tempers the metal and earth yields up the ores. Fire, Air, Water, Earth, and yes, numinous Aether. I am not a shape-shifter like my foster-mother. Instead I shape the subtle structures of matter as well, creating beauty. You think me a god of just one element? Think again.

I was called to bring my forge-hammer to help Zeus birth his beautiful daughter, my half-sister Athena. Ah, she's beautiful, my sister, so beautiful that the first day she stood beside me in the forge my body could not but reach for her. I could not control it, and why should I want to? She was quick to evade me—I get around just fine, but lack speed unless I take the form of searing fire—and wiped off my leaping sperm with, I could see, a flicker of disgust. She has her reasons for keeping to herself and away from the cares of wife and mother, and father Zeus claims much of her time. But she kept the child Mother Gaia bore from my spilled seed. The city of Athens which sprung from that snake-tailed child—which tells us it is truly hers as well, you know her emblem—honors her still, both of us in fact, on the Acropolis (of course they held my torch-race there only this year, as you surely know). My impatient seed was enticed by her beauty and wit, so why should the child so inspired not be thought of as hers?

Athena, she sticks by me. She is smart as a whip, far more than I. Zeus did a better job birthing her, with my help, than Hera did with me, but we might as well be twins. Not the matching kind perhaps, but the kind who come out together, are still bound up with each other. Some so confuse the story they do not know which of us was born first. We work well together. I see the problem, mind the details and create the means with my *technê*. I have the genius of craft, but she has her mother's *mêtis*, the intelligence and craftiness. I have it too: I could not have lived under the sea with my shape-shifting foster-mothers for all that time, and at a formative age, too, without taking it into my young, growing bones, that grew crooked for good reason. And like I said, there is far more to creating what I do than mere skill. But it is Athena who is the master strategist: she mentors gods and mortals alike, and is an inventor too. She understands me better than anyone. How well she deserves the eternal praise of gods and mortals.

Of everything, I love my forge best. I can stay there for eons and never know the time is passing. And it passes efficiently. The assistants I fashioned out of gold, beautiful as nymphs and far more intelligent, are trained to even the finest and most exacting tasks, and they are pleasurable to look at, no less than any of my works. With my twenty bellows and twenty wheeled tripods, the forge blazes with activity. These move like I do, in a circle, rolling crabwise, like the earliest race of beings, who were the most excellent (I'm sorry, but you belong to a degenerate race; I know some of you have figured that out for yourselves). You, who think my twisted legs cripple me; you with your straight legs and feet facing one way only see straight ahead of you, condemned always to move in straight lines. Think about it. And, think about it you should. You have forgotten us gods, most of you. You would do well to remember us now.

Would you really like to know how my bellows and tripods are made? The secret to my golden women, even the secret of Pandora? Ah. There is more than a little magic in what I achieve, and I have many secrets, and no particular need to share them with just anyone.

That's me: Hephaistos, the godly maker. If you want something done and done well, you know where to find me. Put your face into the wind or stare at the sea or mountain; think of the subtle fires burning in your own cells, each a furnace like my own. Feel the flame of your own creative power rising and feel the power that comes to *both* your hands. I am there.

# References

Aeschylus. *Prometheus Bound.* Translated by E. H. Plumptre. Vol. 8, Part 4. The Harvard Classics. New York: P. F. Collier and Son. 1909–14. Bartleby.com, 2001. Accessed March 15, 2004. http://www.bartleby.com/8/4/.

Atchity, Kenneth John. *Homer's Iliad: The Shield of Memory.* Carbondale: Southern Illinois University Press, 1978.

Babalola, Adeboye. "A Portrait of Ogun as Reflected in Ijala Chants." In *Africa's Ogun: Old World and New,* edited by Sandra T. Barnes, 173-198. Bloomington: Indiana University Press, 1997.

Bachelard, Gaston. *Earth and Reveries of Will.* Translated by Kenneth Haltman. Dallas: The Dallas Institute, 2002.

Bachelard, Gaston. *The Psychoanalysis of Fire.* Translated by Alan C. M. Ross. Boston: Beacon, 1964.

Bachofen, J. J. *Myth, Religion and Mother Right: Selected Writings of J. J. Bachofen.* Translated by Ralph Mannheim. Bollingen Series 84. Princeton: Princeton University Press, 1967.

Baldick, Chris. *In Frankenstein's Shadow: Myth, Monstrosity, and Nineteenth-century Writing.* Oxford: Clarendon, 1987.

Barnes, Sandra. T. "The Many Faces of Ogun: Introduction to the First Edition." In *Africa's Ogun: Old World and New,* edited by Sandra T. Barnes, 1-26. Bloomington: Indiana University Press, 1997.

Barnes, Sandra. T., and Paula Girshick Ben-Amos. "Ogun, the Empire Builder." In *Africa's Ogun: Old World and New,* edited by Sandra T. Barnes, 39-64. Bloomington: Indiana University Press, 1997.

Becker, Andrew Sprague. *The Shield of Achilles and the Poetics of Ekphrasis.* London: Rowman and Littlefield, 1995.

Blade Runner Frequently Asked Questions (FAQ). Edited by Murray Chapman. 1992-1995. Accessed August 24, 2003. http://www.faqs.org/faqs/movies/bladerunner-faq/.

Book, James. "Return of a Giant: A fully restored Vulcan— Birmingham's 100-year-old statue—resumes its rightful place in town." About Vulcan (pdf); Vulcan Run 10K and Birmingham Track Club, 2004. Accessed October 30, 2004. http://www.vulcanrun.com/.

Boucher, François. *The Visit of Venus to Vulcan*. 1754. London: Wallace Collection. ArtRussia. Accessed 11. December, 2004. http://www.artrussia.ru/en/picture_rarity/246.

Boucher, François. *Venus Demanding Arms of Vulcan for Aeneas*. 1732. Oil on canvas. 252 x 175 cm. Paris, Musée du Louvre.

Brommer, Frank. *Hephaistos: Der Schmiedegott in der Antiken Kunst*. Mainz am Rhein: Verlag Philipp von Zabern, 1978.

Burkert, Walter. *Greek Religion*. Translated by John Raffan. Cambridge: Harvard University Press, 1985.

Campbell, Joseph. *The Masks of God: Creative Mythology*. New York: Penguin, 1968.

Delcourt, Marie. *Hephaistos ou la lègende du Magicien*. Paris: Les Belles Lettres, 1957.

Detienne, Marcel, and Jean-Pierre Vernant. *Cunning Intelligence in Greek Culture and Society. Translated by Janet Lloyd*. Atlantic Highlands: Humanities, 1978.

Doty, William G. *Mythography: The Study of Myths and Rituals*. 2nd. ed. Tuscaloosa: University of Alabama Press, 2000.

Doty, William G. "What Mythopoetic Means." *Mythosphere* 2.2 (2000): 255-62.

Drewel, Henry John. "Art or Accident: Yoruba Body Artists and Their Deity Ogun." *In Africa's Ogun: Old World and New*, edited by Sandra T. Barnes, 235-64. Bloomington: Indiana University Press, 1997.

Dundes, Alan. "Earth-Diver: Creation of the Mythopoetic Male." In *Sacred Narrative: Readings in the Theory of Myth*, edited by Alan Dundes, 270-294. Berkeley: University of California Press, 1984.

Eisenhower, Dwight D. "Military-Industrial Complex Speech." Public Papers of the Presidents, Dwight D. Eisenhower, 1960, pp. 1035-1040. Accessed October 30, 2004, http://coursesa.matrix.msu.edu/~hst306/documents/indust.html.

Eliade, Mircea. *The Forge and the Crucible: The Origins and Structures of Alchemy*. 2nd. Ed. Chicago: University of Chicago Press, 1956.

Eliade, Mircea. "A New Humanism." In *The Insider/Outsider Problem in the Study of Religion: A Reader*, edited by Russell T. McCutcheon, 95-103. London and New York: Cassell, 1999.

Ellenberger, Henri. *The Discovery of the Unconscious*. New York: Basic Books, 1970.

Farnell, Lewis Richard. *The Cults of the Greek States*. Vol. 5. Chicago: Aegean, 1971.

Forbes, R. J. *Studies in Ancient Technology*. Vol. 8. Leiden: E. J. Brill, 1964.

Ford Motor Company. "Rouge Renovation." 2004. Accessed November 6, 2004. http://www.ford.com/en/goodWorks/environment/cleanerManufacturing/rougeRenovation.htm.

Freedberg, David. *The Power of Images: Studies in the History and Theory of Response*. Chicago: University of Chicago Press, 1989.

Freud, Sigmund. *Leonardo da Vinci and a Memory of His Childhood*. Translated by Alan Tyson. New York: W.W. Norton, 1964.

Freud, Sigmund. "The 'Uncanny.'" In *Writings on Art and Literature*, edited by Werner Hamacher and David E. Wellbery, 193-233. Stanford: Stanford University Press, 1997.

Gablik, Suzi. *The Reenchantment of Art*. New York: Thames and Hudson, 1992.

Gad, Irene. "Hephaestus: Model of New-Age Masculinity." *Quadrant. Journal of the C. G. Jung Foundation for Analytical Psychology*. 2.3 (Fall, 1986): 227-47.

Gantz, Timothy. *Early Greek Myth: A Guide to Literary and Artistic Sources*. Baltimore: The Johns Hopkins University Press, 1993.

Gilliam, Terry, dir. *The Adventures of Baron Munchausen*. (1989; Culver City, CA: Sony Pictures Home Entertainment, 1999), DVD.

Haraway, Donna J. *Simians, Cyborgs, and Women: The Reinvention of Nature*. New York: Routledge, 1991.

Harrison, Jane Ellen. *Prolegomena to the Study of Greek Religion*. Princeton: Princeton University Press, 1991.

Heidegger, Martin. *The Question Concerning Technology and Other Essays*. Translated by William Lovitt. New York: Harper, 1977.

Herodotus. *The Histories*. Translated by A. D. Godley. Cambridge. Harvard University Press. 1920. Perseus Digital Library Project. Edited by Gregory R. Crane. 2004. Tufts University. Accessed September 9, 2004. http://www.perseus.tufts.edu/cgi-bin/ptext?doc=Perseus%3Atext%3A1999.01.0126.

Hesiod. *Theogony, Works and Days, Shield*. Translated by Apostolos N. Athanassakis. Baltimore: Johns Hopkins University Press, 1983.

Hillman, James. *Archetypal Psychology: A Brief Account*. New York: Harper, 1978.

Hillman, James. *A Blue Fire*. Edited by Thomas Moore. New York: Harper, 1989.

Hillman, James. *Healing Fiction*. Woodstock: Spring, 1983.

Hillman, James. "Psychology: Monotheistic or Polytheistic?" In *Working with Images: The Theoretical Base of Archetypal Psychology*, edited by Benjamin Sells, 20-51. Woodstock: Spring, 2000.

Hillman, James. *Re-Visioning Psychology*. New York: Harper, 1976.

Hillman, James. *A Terrible Love of War*. New York: Penguin, 2004.

Hoffmann, E. T. A. "Councilor Krespel." In *E.T.A. Hoffmann Tales*, edited by Victor Lange. The German Library: Vol. 26. New York: Continuum, 1982.

Hoffmann, E. T. A. "The Sandman," 1817. Translated by John Oxenford. *19th Century German Stories*. Edited by Robert Godwin-Jones. 1994. Foreign Lang. Dept., Virginia Commonwealth University. Accessed October 13, 2004. http://www.fln.vcu.edu/hoffmann/sand_e.html.

Homer. *The Iliad*. Translated by Robert Fagles. New York: Penguin, 1990.

Homer. *Iliad. Homeri Opera*. 5 vols. Oxford, Oxford University Press. 1920. Perseus Digital Library Project. Edited by Gregory R. Crane. 2004. Tufts University. Accessed September 12, 2004. http://www.perseus.tufts.edu/cgi-bin/ptext?doc=Perseus%3Atext%3A1999.01.0133.

Homer. *The Odyssey*. Translated by Robert Fagles. New York: Penguin, 1996.

Hyatt, Kenton S. "Creativity Through Intrapersonal Dialog." *Journal of Creative Behavior*. 26.1 (1992): 65-71.

Ingres, Jean Auguste Dominique. *Jupiter and Thetis*. 1811. Oil on canvas. 345 cm. x 257 cm., Aix en-Provence, Musée Granet.

Jung, C. G. *Memories, Dreams, Reflections*. Edited by A. Jaffé. New York: Vintage 1989.

Jung, C. G. *Symbols of Transformation*. Collected Works of C. G. Jung. Rev. by R. F. C. Hull. Translated by H. G. Baynes. Vol. 5. Princeton: Princeton University Press, 1967.

Kac, Eduardo. "Transgenic Art," par. 7. KAC. Accessed November 10, 2004. http://www.ekac.org/transgenic.html.

Kearney, Richard. *The Wake of Imagination*. Minneapolis: University of Minnesota Press, 1988.

Kerényi, Carl. *The Gods of the Greeks*. London: Thames and Hudson, 1951.

LaPlante, Eve. *Seized: Temporal Lobe Epilepsy as Medical, Historical, and Artistic Phenomenon*. New York: Harper, 1993.

Lawal, Babatunde. *The Gèlèdé Spectacle: Art, Gender and Social Harmony in an African Culture*. Seattle: University of Washington Press, 1996.

Lilley, S. *Men, Machines and History*. London: Cobbett, 1948.

Lincoln, Bruce. *Theorizing Myth: Narrative, Ideology, and Scholarship*. Chicago: University of Chicago Press, 1999.

London, Peter. *No More Secondhand Art: Awakening the Artist Within*. Boston: Shambhala, 1989.

Loraux, Nicole. *The Children of Athena: Athenian Ideas About Citizenship and the Division Between the Sexes*. Princeton: Princeton University Press, 1993.

MacCulloch, J. A. *The Celtic and Scandinavian Religions*. Westport: Greenwood, 1973.

Mahony, William K. *The Artful Universe: An Introduction to the Vedic Religious Imagination*. Albany: State University of New York Press, 1998.

Mann, James. *Rise of the Vulcans: The History of Bush's War Cabinet*. New York: Viking, 2004.

MBDC. "Firm Profile." MBDC, LLC. 2001. Accessed November 6, 2004. http://www.mbdc.com/firm_profile.htm.

MBDC. "Monthly Feature: February 2003, A Model for Change." MBDC, LLC. 2003. Accessed November 6, 2004. http://www.mbdc.com/features/feature_feb2003.htm.

"Mike's Pix: Rhythm and Blues 78's" (courtesy of Mike Kredinac). Accessed November 13, 2004. http://www.nugrape.net.

Miller, David L. *Christs: Meditations on Archetypal Images in Christian Theology*. New York: Seabury, 1981.

Miller, David L. *The New Polytheism: Rebirth of the Gods and Goddesses*. New York: Harper, 1974.

Milton, John. *Paradise Lost*. 1667. Renascence Editions. 1992. University of Oregon. Accessed June 9, 2003. http://darkwing.uoregon.edu/~rbear/lost/lost.html.

Monick, Eugene. *Phallos: Sacred Image of the Masculine*. Toronto: Inner City Books, 1987.

Morris, Sarah P. *Daidalos and the Origins of Greek Art*. Princeton: Princeton University Press, 1992.

Motz, Lotte. *The Wise One of the Mountain: Form, Function and Significance of the Subterranean Smith: A Study in Folklore*. Göppingen: Kümmerle Verlag, 1983.

Mumford, Lewis. *Technics and Civilization*. New York: Harcourt, Brace, 1934.

Nadolny, Sten. *The God of Impertinence*. Translated by Breon Mitchell. New York: Viking/Penguin, 1997.

Neumann, Erich. *Origins and History of Consciousness* (Bollingen Series 47). Translated by R. F. C. Hull. Princeton: Princeton University Press, 1970.

Noel, Daniel C. "Veiled Kabir: C. G. Jung's Phallic Self-Image." *Spring*, 1974: 224-42.

Pandilovski, Melentie. "The Ghost of Biotechnology: Art of the Biotech Era," February 2004. The Experimental Art Foundation and Dark Horsey Bookshop, 2004. Accessed November 10, 2004. http://www.eaf.asn.au/essay.htm.

Perseus Vase Catalog. Perseus Digital Library Project. Edited by Gregory R. Crane. 2004. Tufts University. Accessed November 5, 2004, http://www.perseus.tufts.edu/.

Plato. *The Collected Dialogues of Plato*. Eds. Edith Hamilton and Huntington Cairns. Bollingen Series 71. Princeton: Princeton University Press, 1961.

Powell, Michael and Emeric Pressburger, dir. *The Tales of Hoffmann*. 1951; New York: The Criterion Collection, 2005. DVD.

Rollins, Tim, and K.O.S. "Prometheus Bound: A work of art for the web." Dia Art Foundation 1995-2004. Accessed November 6, 2004. http://www.diacenter.org/kos.

Roochnik, David. *Of Art and Wisdom: Plato's Understanding of Techne*. University Park: Pennsylvania State University Press, 1996.

Roochnik, David. *The Tragedy of Reason: Toward a Platonic Conception of Logos*. New York: Routledge, 1990.

Sallis, John. *Being and Logos: The Way of Platonic Dialogue*. 2nd. ed. Atlantic Highlands: Humanities Press International, Inc., 1986.

Scott, Ridley, dir. *Blade Runner*. Director's Cut. 1992; Burbank, CA: Warner Home Video, 1997. DVD.

Scully, Vincent. *The Earth, the Temple, and the Gods*. Rev. ed. New Haven: Yale University Press, 1979.

Shelley, Mary. *Frankenstein, or the Modern Prometheus*. Edited by Johanna M. Smith. Boston: St. Martin's, 1992.

Sinason, Valerie. "Challenged Bodies, Wounded Body Images: Richard III and Hephaestus." In *Splintered Reflections: Images of the Body in Trauma*, edited by Jean Goodwin and Reina Attias, 183-194. New York: Basic Books, 1999.

Slatkin, Laura M. *The Power of Thetis: Allusion and Interpretation in the Iliad*. Berkeley: University of California Press, 1991.

Slochower, Harry. *Mythopoesis: Mythic Patterns in the Literary Classics*. Detroit: Wayne State University Press, 1970.

Smith, Robert C. *The Wounded Jung*. Evanston: Northwestern University Press, 1996.

Solimena, Francesco. *Venus at the Forge of Vulcan*. 1704. Oil on canvas. 205.4 × 153.7 cm. Los Angeles, Getty Center.

Stein, Murray. "Hephaistos: A Pattern of Introversion." In *Facing the Gods*, edited by James Hillman, 35-51. Dallas: Spring. 1980.

Stein, Murray. "The Hephaistos Problem." Rec. July 10, 1993, audiocassette #519. C. G. Jung Institute of Chicago.

Stelarc. "Pingbody; An Internet Actuated and Uploaded Performance." 1996. Accessed November 23, 2004. http://www.stelarc.va.com.au/pingbody/index.html.

*The Homeric Hymns.* Translated by Charles Boer. Woodstock: Spring, 1979.

*The Homeric Hymns.* Translated by Michael Crudden. Oxford World's Classics. Oxford: Oxford University Press, 2001.

TheFreeDictionary.com. "ankulos." Farlex. 2004. Accessed November 20, 2004. http://www.thefreedictionary.com.

Thompson, Stith. *The Folktale.* Berkeley: University of California Press, 1977.

Thompson, Stith. *Motif-Index of Folk-Literature.* Bloomington: Indiana University Press, 1966.

"The Travels And Surprising Adventures Of Baron Munchausen Illustrated With Thirty-Seven Curious Engravings From The Baron's Own Designs And Five Illustrations By G. Cruikshank." 1793. Edited by Marcus Rowland. 2002. Forgotten Futures. Accessed August 24, 2002. http://homepage.ntlworld.com/forgottenfutures/munch/munch.htm.

"Top-Ranked LABS Abstracts 2018." Leonardo Abstracts Service (LABS). Sheila Pinkel, Editor in Chief. Leonardo. Accessed May 21, 2019. https://www.leonardo.info/labs-2018.

Trilling, Lionel. "Art and Neurosis." In *Art and Psychoanalysis,* edited by William Phillips, 502-538. New York: Meridian, 1963.

Velázquez, Diego Rodríguez de Silva y. *Vulcan's Forge.* 1630. Oil on canvas. 223 x 290 cm., Madrid, Museo del Prado.

Vernant, Jean-Pierre, and Pierre Vidal-Naquet. *Myth and Tragedy in Ancient Greece.* Translated by Janet Lloyd. New York: Zone, 1990.

"Vulcan Demo." nZone. © 2003, 2004 NVIDIA® Corporation. Accessed December 6, 2004. http://www.nzone.com/object/nzone_vulcandemo_home.html.

Watkins, Mary. *Invisible Guests: The Development of Imaginal Dialogues*. Dallas: Spring, 2000.

Wtewael, Joachim Anthonisz. *Mars and Venus Surprised by Vulcan*. 1604-1608. Oil on copper. 20.3 x 15.5 cm. Los Angeles, Getty Center.

# Index

Cheryl De Ciantis

of Ouranos, 137; warrior cult
of at Sparta and Corinth, 103
Apollo, 86, 89; and Cyclops, 138,
139; Homeric Hymn to
Pythian Apollo, 93; hostility
toward Telchines, 138;
inspiring poetic vision, 62;
sacred fire of, 156; slayer of
she-dragon child of Hera, 94;
tricked by Hermes' *mêtis*, 60;
witness of Hephaistos'
cuckholding, 5
Apollodorus, 28, 41, 139
archetypal psychology, 161, 172
archetype: archetypes as
interconnected, 26; Hera and
Zeus as archetypal father and
mother, 164; of "mad
scientist", 174; of chthonic
phallic divinities. *See* chthonic
phallic cohort; of
lameness/wounding as magical
initiation, 189; of maker. *See*
maker archetype; split, 21, 26–
27
Arendt, Hannah, 191
Ares, 88, 123; adultery of, 12, 45,
103, 159; and return of
Hephaistos to Olympus, 6,
122; children of with
Aphrodite, 101; ensnared by
Hephaistos, 5, 12, 45, 102, 140
*arête* (excellence): and *technê*,
39–40
Argonauts, 121, 132, 134
Ariadne, 114
Aristophanes, 44
Aristotle, 75–77, 190; ethical
dimension of imagination, 76;
*Nicomachean Ethics*, 37; on
*eidōlopoiountes* (image-
making or "making intellect"),
77; on mathematics, 38
Armitage, Richard, 186
art: and political co-optation of
"master narrative", 182; and

technology, attitudes toward,
19–20, 58; as dialogue, 193; as
false, nondidactic, immoral,
idolatrous, transgressive, 72–
74; as outlet for sexual desire,
163; as process rather than
product, 54, 56–58, 170; as
revealing truth (*alēthea*), 53;
as truth, dealing in essences
(*mythos*), 77; insufficient to
produce "good" for the *polis*,
71; re-mything art and
technology, 192–95; rhetoric
as, 68
Artemis, 89; her chained statue
harming with its gaze, 119;
meteoric iron image of at
Ephesus, 149
artisan, 110, 140, 142, 153, 159,
172; importance of to Athens,
107
artist. *See* also maker archetype;
and creative madness, 16; as
cultural healer, 192; as mere
imitator of Ideal, 71–75; as
revealer of *alēthea* ("truth"),
51; Hephaistos as shadow
aspect of, 162; Kac, Eduardo,
183; of body-scarification
(*olóòlà*), 159; pathologized, 4,
162–72; politically
stigmatized, 4; relationship
with tools of making, 172;
Rollins, Tim and K. O. S.
(Kids of Survival), 192–93;
sexually suspect, 162; Stelarc,
183; suspicion toward, 16, 19;
Telchines makers of first
images of gods, 135; uncanny,
9; wounded, 13, 27, 28, 162–
72, 167
asexual procreation, 5, 97, 104,
105, 136, *See also*
parthenogenesis; and maker
archetype, 130; by Dwarves,
104, 141, 143; Frankenstein,

in iron and steel, 160; of
Achilles, 88; of Apollo, 156;
of Dionysos, 156; of forge, as
magical, 9, 157; of Hector, 87–
88; of Hermes, 156; of Hestia,
156; of Ogun, 158–59;
*phlegma kakon* (deadly
flames), 86, 87, 88, 158; ritual,
expiating womens' "crime" at
Lemnos, 156–57; stolen by
Prometheus, 111; superhuman
fire of shapeshifter Proteus,
88; technological (weapons of
mass destruction), 19, 185;
telluric, 157; *thespiades* (god-
kindled), 87, 88; torch rituals
dedicated to Hephaistos, 156
foot: feet twisted backward, 10,
78, 170; Hephaistos
clubfooted, 43; phallic
symbolism and magic, 10
Forbes, R. J., 32, 34
Ford Motor Company, 194
forge: and sacrifice, 147; as ritual
shrine or sanctuary, 155;
bogey to avert ill-will and
accident, 138, 139; magical
aspects of, 9, 157; volcano as,
78, 157
fostering: as aspect of maker
archetype, 10, 135; Athena
foster-mother of Erichthonios,
79, 104; by chthonic phallic
cohort, 141; by Dwarves, 145;
Eurynome foster-mother of
Hephaistos, 97; infant Zeus
fostered by Dactyls, 135;
infant Zeus fostered by
Kouretes, 135; infant Zeus
fostered by Telchines, 10;
Thetis foster-mother of
Hephaistos, 85, 97, 100
François Krater, 101, 122
*Frankenstein*, 17, 173–74, 180
French Revolution, 24

Freud, Sigmund, 14, 28, 166; on
homosexuality of Leonardo da
Vinci, 168; on pathology and
art, 14; on psychoanalysis as
art, 14; on the uncanny
(*unheimlich*), 175; posthumous
analysis of Leonardo da Vinci,
162–63; split from C. G. Jung,
14
Gablik, Suzi, 192
Gad, Irene, 163
Gaia, 10, 104–5; and castration of
Ouranos, 137, 144;
fecundating power of Earth,
94, 95, 105; mother of
Cyclops, 138; mother of
Erichthonios, 5, 79, 104, 109;
mother of first generation of
gods, 95, 98, 136
gait, 4; deformed gait as analogy
for socially illegitimate
thought and speech, 44;
dwarves in folklore feet
twisted backward, 10;
Hephaistos feet twisted
backward, 10, 78, 170;
Hermes' reversed gait as
emblematic of *mêtis*, 44; of
Hephaistos as emblematic of
*mêtis*, 43, 71, 78, 102; skewed
or crippled, symbolizing
possession of *mêtis*, 43–44;
straight gait as analogy for
character of legitimate ruling
class, 44; straight gait as
metaphor for speech of reason
(*logos*), 70
Ganesha, 97
Gantz, Timothy, 47, 48, 101, 108,
112, 137
Gaskell, Elizabeth, 181
giant, 142, 150; frost-giant Ymir,
143
*Gilgamesh*, 151
Gilliam, Terry, 18, 184
Gimbutas, Marija, 49

Temporal Lobe Syndrome (TLS),
16
Tethys. *See also* Thetis; Oceanic
powers; as origin of all things,
100; wife of Okeanos, 95
Thebes: cult site of chthonic
phallic divinities, 8, 131, 132,
133; cult site of Demeter
Kabeiraia, 133
Theophilus, 143
*theōria* (theory): as attentiveness,
51–52
Thetis, 7, 97–100, *See also*
Oceanic powers; "mortal pain"
of, 98; and Zeus, 98–100; as
analogue of Lady Athirat in
Ugaritic myths of "wise ones",
153; as primordial creative
power, 100, 121; bound in
marriage to mortal Peleus, 98,
100, 120; epithet "Silver-
Footed", 94, 97; foster-mother
of Hephaistos, 84, 85, 97, 98,
165; identified with Tethys
wife of Okeanus, origin of all
things, 100; in the workshop of
Hephaistos, 83–85; *mêtis* of,
121; mother of Achilles, 98;
persuasive *logos* of, 80, 99;
prophesied to bear son to
overthrow Zeus, 100, 112;
rescuer of Dionysos, 100;
rescuer of Hephaistos, 6, 100;
rescuer of Zeus from binding,
99, 120; sea-nymph, daughter
of Nereus, 93, 98
Thompson, Stith, 146
thunderbolt: etymological
connection with stoneworking,
50; forged by Cyclops, 138,
150; forged by Dwarves, 148;
initiating heaven-earth
hierogamy, 150; meteoritic
iron as symbolic of, 150; of
Zeus forged by Telchines, 138

Titans, 112, 120; bound in
Tartarus, 99; Prometheus, 111
trickster: Hephaistos, 6, 112;
Hermes, 59; Loki, 146;
possessor of *mêtis*, 42;
Prometheus, 111, 112
truth. *See alēthea*
Typhaon, 94, 95
uncanny, 9; artistic
representations of life as, 16;
Freud on the uncanny
(*unheimlich*), 175; Santa
Claus, 177
unconscious, 14
uroborus, 129
Vac, 147
values: "value-neutrality" of
technology, 4, 36–37; imposed
on others by those holding
power, 70
Van Gogh, Vincent, 16
Vasari, Giorgio, 162
Vedas, 148; as magical songs,
147
Velázquez, Diego, 12
Venus, 13
Vernant, Jean-Pierre, 27, 28, 40–
46, 121, 132
Vidal-Naquet, Pierre, 28, 43–44
Virgil: *Aeneid*, 12; *Eclogues*, 138
volcano: forge of Hephaistos, 78,
157; Hermes bound in, 185
Vulcan, 18, 158; in Renaissance
and Baroque art, 11–13;
popular culture associations of,
16; statue of in Birmingham,
AL, 186
Wagner, Richard, 188
Watkins, Mary, 29, 161, 190–91
Wayland, 188–89, *See also*
blacksmith divinities
Wieland, 10, 156, *See also*
blacksmith divinities
wine, 122, 132; mythic effects of,
165

# Acknowledgements

FROM 1999 through 2003, while I and my cohort were completing our Doctoral course work, we at Pacifica had access to an extraordinary group of scholars, many emeritus from signal institutions and a small number who affiliated directly with C.G. Jung and the early circle of Eranos scholars, notably David L. Miller. Dr. Miller and Dr. Dennis Slattery exemplify the art of querying the Text to explore its depth psychological core—though as Heraclitus reminds us, we can never plumb the limits of the breadth and depth of the well of meaning that is our human endowment. The late James Hillman served on the Board and presented an annual lecture to students, and his lasting influence is evidenced in the gift of his library to Pacifica. Joseph Campbell aided in Pacifica's founding, and I spent many hours searching out the tiny, penciled marginalia in the books he gifted to Pacifica, which houses his personal library as part of the The Joseph Campbell Collection in the remarkable Opus Archives, which also houses the Marija Gimbutas Library.

Like any work of its kind, this dissertation could not have been made without helping heads, hands, and hearts. My committee Chair, Dr. Ginette Paris, captured my imagination in our first Pacifica class on Greek mythology when instead of describing it she suddenly came around the lectern and crab-walked across the floor in Hephaistos' startling, circular, clubfooted gait. As my Advisor, Dr. Christine Downing's profound scholarship and her gentle butexqisitely incisive questions set an example I will always aspire to. My External Reader, Dr. Gilles Zenon Maheu, kept my feet on the ground with regard to delving sufficiently into the origins and early developments of technology as we know it.

My dear friend and mentor, artist, educator and activist Sheila Pinkel has been an inspiring presence in my life for decades and was the reader I imagined looking over my shoulder as I wrote. I was similarly lucky to find in the late Diane Downey another extraordinary mentor and friend who inducted me into the mysteries of organizational life and fostered my entry into the world of organizational creativity with its archetypal frustrations, perennial dysfunctions and boundless opportunities. I have had the help and love of not just one but two groups of extraordinarily talented and wise women. To my faithful and longstanding women's group, Rebecca Henson, Joyce Richman, Donna Riechmann, Martha Tilyard, and the late Susan Allen, as well as my original dissertation support group, who morphed into the Myth Sisters, Victoria Barnes, Maila Davenport, the late Jane Estelle, Jane Feldman, Ann Bowie Maxwell, and Priscilla Taylor, I offer my thanks from the very bottom of my heart. I offer thanks too to another wildwoman and iconoclastic agent of inspiration, Annette Simmons. Ginger Grant is in a class by herself and I'm the luckier to know her. It has been my good fortune to know wonderfully creative men who remain dear friends, among them Mahmood Karimi-Hakak, Chris Musselwhite, and mi caro amico Robert Burnside. Best of all, my soul mate in this life, Kenton Hyatt. He is my live-in professor, coach, reader, affectionate supporter, and a lot of other good things I leave to the imagination of the reader. I already knew how generous and loving he is, but I had to go to graduate school to even begin to appreciate quite how marvelously intelligent he is, and he continues to surprise me. I have no idea what it would have been like to tackle a project like this without him. With him, it was a delight.

And, I thank Hephaistos, who showed up, and to whom I pray (risking the further consequences of such hubris), "Stick around! We have a lot more Making to do."

ABOUT THE AUTHOR

CHERYL DE CIANTIS is an artist, writer, storyteller and mythologist with an interest in the intersection of art, technology and culture. A Senior Faculty emerita and former Director of the European Campus of the Center for Creative Leadership, she has facilitated, coached and mentored creative leaders for long enough to be reckoned an entry-level elder. Cheryl earned her Ph.D. in Mythological Studies with Emphasis in Depth Psychology at Pacifica Graduate Institute, which also houses the Joseph Campbell Library. She lives with her work- and life-partner, Kenton Hyatt, in Tucson, Arizona. You can visit her artist website at cheryldeciantis.com.

www.ingramcontent.com/pod-product-compliance
Lightning Source LLC
Chambersburg PA
CBHW031246090426
42742CB00007B/340